T0297847

An Introduction to the
Analysis of Algorithms

An Introduction to the
Analysis of Algorithms

Michael Soltys
McMaster University, Canada

World Scientific

NEW JERSEY · LONDON · SINGAPORE · BEIJING · SHANGHAI · HONG KONG · TAIPEI · CHENNAI

Published by

World Scientific Publishing Co. Pte. Ltd.
5 Toh Tuck Link, Singapore 596224
USA office: 27 Warren Street, Suite 401-402, Hackensack, NJ 07601
UK office: 57 Shelton Street, Covent Garden, London WC2H 9HE

British Library Cataloguing-in-Publication Data
A catalogue record for this book is available from the British Library.

AN INTRODUCTION TO THE ANALYSIS OF ALGORITHMS

ISBN-13 978-981-4271-40-0
ISBN-10 981-4271-40-3

Printed in Singapore by Mainland Press Pte Ltd.

To my family

Preface

This book is an introduction to the analysis of algorithms, from the point of view of proving algorithm correctness. Our theme is the following: how do we argue mathematically, without a burden of excessive formalism, that a given algorithm does what it is supposed to do? And why is this important? In the words of C.A.R. Hoare:

> As far as the fundamental science is concerned, we still certainly do not know how to prove programs correct. We need a lot of steady progress in this area, which one can foresee, and a lot of breakthroughs where people suddenly find there's a simple way to do something that everybody hitherto has thought to be far too difficult[1].

Software engineers know many examples of things going terribly wrong because of program errors; their particular favorites are the following two[2]. The blackout in the American North-East during the summer of 2003 was due to a software bug in an energy management system; an alarm that should have been triggered never went off, leading to a chain of events that climaxed in a cascading blackout. The Ariane 5, flight 501, the maiden flight of the rocket in June 4, 1996, ended with an explosion 40 seconds into the flight; this $500 million loss was caused by an overflow in the conversion from a 64-bit floating point number to a 16-bit signed integer.

While the goal of absolute certainty in program correctness is elusive, we can develop methods and techniques for reducing errors. The aim of this book is modest: we want to present an introduction to the analysis of algorithms—the "ideas" behind programs, and show how to prove their correctness.

---

[1] From *An Interview with C.A.R. Hoare*, in [Shustek (2009)].

[2] These two examples come from [van Vliet (2000)], where many more instances of spectacular failures may be found.

The algorithm may be correct, but the implementation itself might be flawed. Some syntactical errors in the program implementation may be uncovered by a compiler or translator—which in turn could also be buggy—but there might be other hidden errors. The hardware itself might be faulty; the libraries on which the program relies at run time might be unreliable, etc. It is the task of the software engineer to write code that works given such a delicate environment, prone to errors. Finally, the algorithmic content of a piece of software might be very small; the majority of the lines of code could be the "menial" task of interface programming. Thus, the ability to argue correctly about the soundness of an algorithm is only one of many facets of the task at hand, yet an important one, if only for the pedagogical reason of learning to argue rigorously about algorithms.

We begin this book with a chapter of preliminaries, containing the key ideas of induction and invariance, and the framework of pre/post-conditions and loop invariants.

The remaining, purely algorithmic, contents of the book are as follows. We present three standard algorithm design techniques in eponymous chapters: greedy algorithms, dynamic programming and the divide and conquer paradigm. We are concerned with correctness of algorithms, rather than, say, efficiency or the underlying data structures. For example, in the chapter on the greedy paradigm we explore in depth the idea of a *promising partial solution*, a powerful technique for proving the correctness of greedy algorithms. We also include online algorithms and the idea of *competitive analysis*, and the last chapter is an introduction to randomized algorithms, with a section on cryptography.

The intended audience for this book consists of undergraduate students in computer science and mathematics. The book is very self-contained: the first chapter, Preliminaries, reviews induction and the invariance principle. It also introduces the aforementioned ideas of pre/post-conditions, loop invariants and termination—in other words, it sets the mathematical stage for the rest of the book. Not much mathematics is assumed (besides some tame forays into linear algebra and number theory), but a certain penchant for discrete mathematics is considered helpful.

Algorithms solve problems, and many of the problems in this book fall under the category of *optimization problems*, whether cost minimization, such as Kruskal's algorithm for computing minimum cost spanning trees—section 2.1, or profit maximization, such as selecting the most profitable subset of activities—section 4.5.

This book draws on many sources. First of all, [Cormen *et al.* (2001)] is a fantastic reference for anyone who is learning algorithms. I have also used as reference the elegantly written [Kleinberg and Tardos (2006)]. A classic in the field is [Knuth (1997)], and I base my presentation of online algorithms on the material in [Borodin and El-Yaniv (1998)]. I have learned greedy algorithms and dynamic programming from Stephen A. Cook at the University of Toronto. Appendix B, on relations, is based on hand-written lectures slides of Ryszard Janicki.

No book on algorithms is complete without a short introduction to the "big-Oh" notation. We say that $g(n) \in O(f(n))$ if there exist constants c, n_0 such that for all $n \geq n_0$, $g(n) \leq cf(n)$, and the *little-oh* notation, $g(n) \in o(f(n))$, which denotes that $\lim_{n \to \infty} g(n)/f(n) = 0$. We also say that $g(n) \in \Omega(f(n))$ if there exist constants c, n_0 such that for all $n \geq n_0$, $g(n) \geq cf(n)$. Finally, we say that $g(n) \in \Theta(f(n))$ if it is the case that both $g(n) \in O(f(n))$ and $g(n) \in \Omega(f(n))$.

The ubiquitous *floor* and *ceil* functions are defined, respectively, as follows: $\lfloor x \rfloor = \max\{n \in \mathbb{Z} | n \leq x\}$ and $\lceil x \rceil = \min\{n \in \mathbb{Z} | n \geq x\}$.

Contents

Preface   vii

1. Preliminaries   1

    1.1   Induction   1

    1.2   Invariance   4

    1.3   Correctness of algorithms   5

          1.3.1   Division algorithm   6

          1.3.2   Euclid's algorithm   7

          1.3.3   Palindromes algorithm   8

          1.3.4   Further examples   9

          1.3.5   Recursion and fixed points 12

    1.4   Stable marriage 14

    1.5   Answers to selected problems 17

    1.6   Notes 27

2. Greedy Algorithms   29

    2.1   Minimum cost spanning trees 29

    2.2   Jobs with deadlines and profits 34

    2.3   Further examples and problems 36

          2.3.1   Make change 36

          2.3.2   Maximum weight matching 37

          2.3.3   Shortest path 38

    2.4   Answers to selected problems 38

    2.5   Notes 42

3.　Divide and Conquer   43

　　3.1　Mergesort . . . .　44
　　3.2　Multiplying numbers in binary . . . .　45
　　3.3　Savitch's algorithm . . . .　48
　　3.4　Answers to selected problems . . . .　49
　　3.5　Notes . . . .　50

4.　Dynamic Programming   51

　　4.1　Longest monotone subsequence problem . . . .　51
　　4.2　All pairs shortest path problem . . . .　52
　　4.3　Simple knapsack problem . . . .　54
　　　　4.3.1　Dispersed knapsack . . . .　58
　　4.4　General knapsack problem . . . .　58
　　4.5　Activity selection problem . . . .　59
　　4.6　Jobs with deadlines, durations and profits . . . .　62
　　4.7　Further examples and problems . . . .　63
　　　　4.7.1　Consecutive subsequence sum problem . . . . . .　63
　　　　4.7.2　Context free grammars . . . .　64
　　4.8　Answers to selected problems . . . .　67
　　4.9　Notes . . . .　71

5.　Online Algorithms   73

　　5.1　List accessing problem . . . .　74
　　5.2　Paging . . . .　78
　　　　5.2.1　Demand paging . . . .　79
　　　　5.2.2　FIFO . . . .　82
　　　　5.2.3　LRU . . . .　82
　　　　5.2.4　Marking algorithms . . . .　86
　　　　5.2.5　FWF . . . .　87
　　　　5.2.6　LFD . . . .　88
　　5.3　Answers to selected problems . . . .　92
　　5.4　Notes . . . .　94

6.　Randomized Algorithms   95

　　6.1　Perfect matching . . . .　96
　　6.2　Pattern matching . . . .　100
　　6.3　Primality testing . . . .　101
　　6.4　Public key cryptography . . . .　105

	6.4.1	Diffie-Hellman key exchange	106
	6.4.2	ElGamal	107
	6.4.3	RSA	107
6.5		Answers to selected problems	109
6.6		Notes	111

Appendix A   Number Theory and Group Theory   113

| A.1 | Answers to selected problems | 118 |
| A.2 | Notes | 118 |

Appendix B   Relations   119

B.1	Closure	120
B.2	Equivalence relation	121
B.3	Partial orders	123
B.4	Lattices	124
B.5	Fixed point theory	125
B.6	Answers to selected problems	128
B.7	Notes	128

Bibliography   129

Index   131

Chapter 1

Preliminaries

1.1 Induction

Let $\mathbb{N} = \{0, 1, 2, \ldots\}$ be the set of natural numbers. Suppose that S is a subset of $\mathbb{N}$ with the following two properties: first $0 \in S$, and second, whenever $n \in S$, then $n + 1 \in S$ as well. Then, invoking the *Induction Principle (IP)* we can conclude that $S = \mathbb{N}$.

We shall use the IP with a more convenient notation; let P be a property of natural numbers, in other words, P is a unary relation such that $P(i)$ is either true or false. The relation P may be identified with a set S_P in the obvious way, i.e., $i \in S_P$ iff $P(i)$ is true. For example, if P is the property of being prime, then $P(2)$ and $P(3)$ are true, but $P(6)$ is false, and $S_P = \{2, 3, 5, 7, 11, \ldots\}$. Using this notation the IP may be stated as:

$$[P(0) \wedge \forall n(P(n) \rightarrow P(n+1))] \rightarrow \forall m P(m), \qquad (1.1)$$

for any (unary) relation P over $\mathbb{N}$. In practice, we use (1.1) as follows: first we prove that $P(0)$ holds (this is the *basis case*). Then we show that $\forall n(P(n) \rightarrow P(n+1))$ (this is the *induction step*). Finally, using (1.1) and *modus ponens*, we conclude that $\forall m P(m)$.

As an example, let P be the assertion "the sum of the first i odd numbers equals i^2." We follow the convention that the sum of an empty set of numbers is zero; thus $P(0)$ holds as the set of the first zero odd numbers is an empty set. $P(3)$ is also true as $1 + 3 + 5 = 9 = 3^2$. We want to show that in fact $\forall m P(m)$ (i.e., P is always true, and so $S_P = \mathbb{N}$).

We use induction. The basis case is $P(0)$ and we already showed that it holds. Suppose now that the assertion holds for n, i.e., the sum of the first n odd numbers is n^2, i.e., $1 + 3 + 5 + \cdots + (2n-1) = n^2$ (this is our *inductive hypothesis* or *inductive assumption*). Consider the sum of the first $(n + 1)$

1

odd numbers,

$$\boxed{1 + 3 + 5 + \cdots + (2n-1)} + (2n+1) = \boxed{n^2} + (2n+1) = (n+1)^2,$$

and so we just proved the induction step, and by IP we have $\forall m P(m)$.

Problem 1.1. *Prove that* $1 + \sum_{j=0}^{i} 2^j = 2^{i+1}$.

Sometimes it is convenient to start our induction higher than at 0. We have the following generalized induction principle:

$$[P(k) \wedge \forall n(P(n) \rightarrow P(n+1))] \rightarrow (\forall m \geq k)P(m), \qquad (1.2)$$

for any predicate P and any number k. Note that (1.2) follows easily from (1.1) if we simply let $P'(i)$ be $P(i+k)$, and do the usual induction on the predicate $P'(i)$.

Problem 1.2. *Use induction to prove that for $n \geq 1$,*

$$1^3 + 2^3 + 3^3 + \cdots + n^3 = (1 + 2 + 3 + \cdots + n)^2.$$

Problem 1.3. *For every $n \geq 1$, consider a square of size $2^n \times 2^n$ where one square is missing. Show that the resulting square can be filled with "L" shapes—that is, with clusters of three squares, where the three squares do not form a line.*

Problem 1.4. *In the generalized IP (1.2) we can replace the induction step $\forall n(P(n) \rightarrow P(n+1))$ with $(\forall n \geq k)(P(n) \rightarrow P(n+1))$. Explain why both versions effectively yield the same principle.*

Problem 1.5. *The Fibonacci sequence is defined as follows: $f_0 = 0$ and $f_1 = 1$ and $f_{i+2} = f_{i+1} + f_i$, $i \geq 0$. Prove that for all $n \geq 1$ we have:*

$$\begin{pmatrix} 1 & 1 \\ 1 & 0 \end{pmatrix}^n = \begin{pmatrix} f_{n+1} & f_n \\ f_n & f_{n-1} \end{pmatrix}.$$

Problem 1.6. *Prove the following: if m divides n, then f_m divides f_n, i.e., $m|n$ implies $f_m|f_n$.*

The *Complete Induction Principle* (CIP) is just like IP except that in the induction step we show that if $P(i)$ holds for all $i \leq n$, then $P(n+1)$ also holds, i.e., the induction step is now $\forall n((\forall i \leq n)P(i) \rightarrow P(n+1))$.

Problem 1.7. *Use the CIP to prove that every number (in $\mathbb{N}$) greater than 1 may be written as a product of one or more prime numbers.*

Problem 1.8. *Suppose that we have a (Swiss) chocolate bar consisting of a number of squares arranged in a rectangular pattern. Our task is to split the bar into small squares (always breaking along the lines between the squares) with a minimum number of breaks. How many breaks will it take? Make an educated guess, and prove it by induction.*

The *Least Number Principle (LNP)* says that every non-empty subset of the natural numbers must have a least element. A direct consequence of the LNP is that every decreasing non-negative sequence of integers must terminate; that is, if $R = \{r_1, r_2, r_3, \ldots\} \subseteq \mathbb{N}$ where $r_i > r_{i+1}$ for all i, then R is a *finite* subset of $\mathbb{N}$. We are going to be using the LNP to show termination of algorithms.

Problem 1.9. *Show that IP, CIP, and LNP are equivalent principles.*

There are three standard ways to list the nodes of a binary tree. We present them below, together with a recursive procedure that lists the nodes according to each scheme.

Infix: left sub-tree, root, right sub-tree.
Prefix: root, left sub-tree, right sub-tree.
Postfix: left sub-tree, right sub-tree, root.
See the example in figure 1.1.

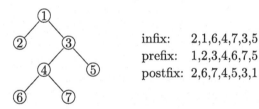

infix: 2,1,6,4,7,3,5
prefix: 1,2,3,4,6,7,5
postfix: 2,6,7,4,5,3,1

Fig. 1.1   A binary tree with the corresponding representations.

Note that some authors use a different name for infix, prefix, and postfix; they call it inorder, preorder, and postorder, respectively.

Problem 1.10. *Show that given any two representations we can obtain from them the third one, or, put another way, from any two representations we can reconstruct the tree. Show, using induction, that your reconstruction is correct. Then show that having just one representation is not enough.*

1.2　Invariance

The *Invariance Technique (IT)* is a method for proving assertions about the outcomes of procedures. The IT identifies some property that remains true throughout the execution of a procedure. Then, once the procedure terminates, we use this property to prove assertions about the output.

Fig. 1.2　An 8 × 8 board.

As an example, consider an 8 × 8 board from which two squares from opposing corners have been removed (see figure 1.2). The area of the board is $64 - 2 = 62$ squares. Now suppose that we have 31 dominoes of size 1×2. We want to show that the board *cannot* be covered by them.

Verifying this by *brute force* (that is, examining all possible coverings) is an extremely laborious job. However, using IT we argue as follows: color the squares as a chess board. Each domino, covering two adjacent squares, covers 1 white and 1 black square, and, hence, each placement covers as many white squares as it covers black squares. Note that the number of white squares and the number of black squares differ by 2—opposite corners lying on the same diagonal have the same color—and, hence, no placement of domino's yields a cover; done!

More formally, we place the dominoes one by one on the board, any way we want. The invariant is that after placing each new domino, the number of covered white squares is the same as the number of covered black squares. We prove that this *is* an invariant by induction on the number of placed dominoes. The basis case is when zero dominoes have been placed (so zero black and zero white squares are covered). In the induction step, we add one more domino which, no matter how we place it, covers one white and one black square, thus maintaining the property. At the end, when we are done placing dominoes, we would have to have as many white squares as black squares covered, which is not possible due to the nature of the coloring of the board (i.e., the number of black and whites squares is not the same). Note that this argument extends easily to the $n \times n$ board.

Problem 1.11. *Let n be an odd number, and suppose that we have the set $\{1, 2, \ldots, 2n\}$. We pick any two numbers a, b in the set, delete them from the set, and replace them with $|a - b|$. Continue repeating this until just one number remains in the set; show that this remaining number must be odd.*

The next three problems have the common theme of social gatherings. We always assume that relations of likes and dislikes, of being an enemy or a friend, are reflexive relations: that is, if a likes b, then b also likes a, etc. See appendix B for background on relations—reflexive relations are defined on page 119.

Problem 1.12. *At a country club, each member dislikes at most three other members. There are two tennis courts; show that each member can be assigned to one of the two courts in such a way that at most one person they dislike is also playing on the same court.*

Problem 1.13. *You are hosting a dinner party where $2n$ people are going to be sitting at a round table. As it happens in any social clique, animosities are rife, but you know that everyone sitting at the table dislikes at most $(n-1)$ people; show that you can make sitting arrangements so that nobody sits next to someone they dislike.*

Problem 1.14. *Handshakes are exchanged at a meeting. We call a person an* odd *person if he has exchanged an odd number of handshakes. Show that, at any moment, there is an even number of odd persons.*

1.3 Correctness of algorithms

How can we prove that an algorithm is correct[1]? We make two assertions, called the *pre-condition* and the *post-condition*; by correctness we mean that whenever the pre-condition holds *before* the algorithm executes, the post-condition will hold *after* it executes. By *termination* we mean that whenever the pre-condition holds, the algorithm will stop running after finitely many steps. Correctness without termination is called *partial correctness*, and *correctness* per se is partial correctness *with* termination.

---

[1] A wonderful introduction to this topic can be found in [Harel (1987)], in chapter 5, "The correctness of algorithms, or getting it done right."

A fundamental notion in the analysis of algorithms is that of a *loop invariant*; it is an assertion that stays true after each execution of a "while" (or "for") loop. Coming up with the right assertion, and proving it, is a creative endeavor.

Once the loop invariant has been shown to hold, it is used for proving partial correctness of the algorithm. So the criterion for selecting a loop invariant is that it helps in proving the post-condition. In general many different loop invariants (and for that matter pre and post-conditions) may yield a desirable proof of correctness; the art of the analysis of algorithms consists in selecting them judiciously. We usually need induction to prove that a chosen loop invariant holds after each iteration of a loop, and usually we also need the pre-condition as an assumption in this proof.

An implicit pre-condition of all the algorithms in this section is that the numbers are in $\mathbb{Z}$.

1.3.1 *Division algorithm*

We analyze the algorithm for integer division, algorithm 1.1. Note that the q and r returned by the division algorithm are usually denoted as $\mathrm{div}(x, y)$ (the *quotient*) and $\mathrm{rem}(x, y)$ (the *remainder*), respectively.

Algorithm 1.1 Division

Pre-condition: $x \geq 0 \wedge y > 0$

1: $q \leftarrow 0$
2: $r \leftarrow x$
3: **while** $y \leq r$ **do**
4:      $r \leftarrow r - y$
5:      $q \leftarrow q + 1$
6: **end while**
7: **return** q, r

Post-condition: $x = (q \cdot y) + r \wedge 0 \leq r < y$

We propose the following assertion as the loop invariant:

$$x = (q \cdot y) + r \wedge r \geq 0. \tag{1.3}$$

We show that (1.3) holds after each iteration of the loop. Basis case (i.e., zero iterations of the loop—we are just before line 3 of the algorithm): $q = 0, r = x$, so $x = (q \cdot y) + r$ and since $x \geq 0$ and $r = x$, $r \geq 0$.

Induction step: suppose $x = (q \cdot y) + r \wedge r \geq 0$ and we go once more through the loop, and let q', r' be the new values of q, r, respectively (computed in lines 4 and 5 of the algorithm). Since we executed the loop one more time it follows that $y \leq r$ (this is the condition checked for in line 3 of the algorithm), and since $r' = r - y$, we have that $r' \geq 0$. Thus,

$$x = (q \cdot y) + r = ((q + 1) \cdot y) + (r - y) = (q' \cdot y) + r',$$

and so q', r' still satisfy the loop invariant (1.3).

Now we use the loop invariant to show that (if the algorithm terminates) the post-condition of the division algorithm holds, *if* the pre-condition holds. This is very easy in this case since the loop ends when it is no longer true that $y \leq r$, i.e., when it is true that $r < y$. On the other hand, (1.3) holds after each iteration, and in particular the last iteration. Putting together (1.3) and $r < y$ we get our post-condition, and hence partial correctness.

To show termination we use the least number principle (LNP). We need to relate some non-negative monotone decreasing sequence to the algorithm; just consider $r_0, r_1, r_2, \ldots$, where $r_0 = x$, and r_i is the value of r after the i-th iteration. Note that $r_{i+1} = r_i - y$. First, $r_i \geq 0$, because the algorithm enters the while loop only if $y \leq r$, and second, $r_{i+1} < r_i$, since $y > 0$. By LNP such a sequence "cannot go on for ever," (in the sense that the set $\{r_i | i = 0, 1, 2, \ldots\}$ is a subset of the natural numbers, and so it has a least element), and so the algorithm must terminate.

Thus we have shown full correctness of the division algorithm.

1.3.2 *Euclid's algorithm*

Given two positive integers a and b, their *greatest common divisor*, denoted as $\gcd(a, b)$, is the largest positive integer that divides them both. Euclid's algorithm, presented as algorithm 1.2, is a procedure for finding the greatest common divisor of two numbers. It is one of the oldest know algorithms— it appeared in Euclid's *Elements* (Book 7, Propositions 1 and 2) around 300 BC.

Note that to compute $\text{rem}(n, m)$ in lines 1 and 3 of Euclid's algorithm we need to use algorithm 1.1 (the division algorithm) as a subroutine; this is a typical "composition" of algorithms. Also note that lines 1 and 3 are executed from left to right, so in particular in line 3 we first do $m \leftarrow n$, then $n \leftarrow r$ and finally $r \leftarrow \text{rem}(m, n)$. This is important for the algorithm to work correctly.

Algorithm 1.2 Euclid

Pre-condition: $a > 0 \wedge b > 0$

1: $m \leftarrow a$; $n \leftarrow b$; $r \leftarrow \mathrm{rem}(m, n)$
2: **while** $(r > 0)$ **do**
3:        $m \leftarrow n$; $n \leftarrow r$; $r \leftarrow \mathrm{rem}(m, n)$
4: **end while**
5: **return** n

Post-condition: $n = \gcd(a, b)$

To prove the correctness of Euclid's algorithm we are going to show that after each iteration of the while loop the following assertion holds:

$$\gcd(m, n) = \gcd(a, b), \tag{1.4}$$

that is, (1.4) is our loop invariant. We prove this by induction on the number of iterations. Basis case: after zero iterations (i.e., just before the while loop starts—so after executing line 1 and before executing line 2) we have that $m = a$ and $n = b$, so (1.4) holds trivially.

For the induction step, suppose that $\gcd(a, b) = \gcd(m, n)$, and we go through the loop one more time, yielding m', n'. We want to show that $\gcd(m, n) = \gcd(m', n')$. Note that from line 3 of the algorithm we see that $m' = n, n' = r = \mathrm{rem}(m, n)$. In other words, it is enough to prove that in general $\gcd(m, n) = \gcd(n, \mathrm{rem}(m, n))$.

Problem 1.15. *Show that for all $m, n > 0$, $\gcd(m, n) = \gcd(n, \mathrm{rem}(m, n))$.*

Now the correctness of Euclid's algorithm follows from (1.4), since the algorithm stops when $r = \mathrm{rem}(m, n) = 0$, so $m = q \cdot n$, and so $\gcd(m, n) = n$.

Problem 1.16. *Show that Euclid's algorithm terminates.*

Problem 1.17. *Do you have any ideas how to speed-up Euclid's algorithm?*

Problem 1.18. *Given integers n, m, and $d = \gcd(n, m)$, it is possible to compute integers a, b such that $an + bm = d$. The algorithm for computing a, b (besides also computing the $\gcd(n, m)$) is called the extended Euclid's algorithm. Design this algorithm and prove its correctness.*

1.3.3   *Palindromes algorithm*

Algorithm 1.3 tests strings for *palindromes*, which are strings that read the same backwards as forwards, for example, `madamimadam` or `racecar`.

Algorithm 1.3 Palindromes

Pre-condition: $n \geq 1 \wedge A[1 \ldots n]$ is a character array

1: $i \leftarrow 1$
2: **while** $(i \leq \lfloor \frac{n}{2} \rfloor)$ **do**
3:     **if** $(A[i] \neq A[n-i+1])$ **then**
4:         **return** F
5:         $i \leftarrow i+1$
6:     **end if**
7: **end while**
8: **return** T

Post-condition: return T iff A is a palindrome

Let the loop invariant be: after the k-th iteration, $i = k+1$ and for all j such that $1 \leq j \leq k$, $A[j] = A[n-j+1]$. We prove that the loop invariant holds by induction on k. Basis case: before any iterations take place (i.e., after zero iterations), there are no j's such that $1 \leq j \leq 0$, so the second part of the loop invariant is (vacuously) true. The first part of the loop invariant holds since i is initially set to 1.

Induction step: we know that after k iterations, $A[j] = A[n-j+1]$ for all $1 \leq j \leq k$; after one more iteration we know that $A[k+1] = A[n-(k+1)+1]$, so the statement follows for all $1 \leq j \leq k+1$. This proves the loop invariant.

Problem 1.19. *Using the loop invariant argue the partial correctness of the palindromes algorithm. Show that the algorithm for palindromes always terminates.*

1.3.4 Further examples

Problem 1.20. *Give an algorithm which on the input "a positive integer n," outputs "yes" if $n = 2^k$ (i.e., n is a power of 2), and "no" otherwise. Prove that your algorithm is correct.*

Problem 1.21. *What does algorithm 1.4 compute? Prove your claim.*

Problem 1.22. *What does algorithm 1.5 compute? Assume that a, b are positive integers (i.e., assume that the pre-condition is that $a, b > 0$). For which starting a, b does this algorithm terminate? In how many steps does it terminate, if it does terminate?*

Algorithm 1.4 Problem 1.21

1: $x \leftarrow m \; ; \; y \leftarrow n \; ; \; z \leftarrow 0$
2: **while** $(x \neq 0)$ **do**
3:      **if** $(\mathrm{rem}(x, 2) = 1)$ **then**
4:          $z \leftarrow z + y$
5:      **end if**
6:      $x \leftarrow \mathrm{div}(x, 2)$
7:      $y \leftarrow y \cdot 2$
8: **end while**
9: **return** z

Algorithm 1.5 Problem 1.22

1: **while** $(a > 0)$ **do**
2:      **if** $(a < b)$ **then**
3:          $(a, b) \leftarrow (2a, b - a)$
4:      **else**
5:          $(a, b) \leftarrow (a - b, 2b)$
6:      **end if**
7: **end while**

Problem 1.23. *The following problem requires some linear algebra[2]. Let $\{v_1, v_2, \ldots, v_n\}$ be a basis for a vectors space $V \subseteq \mathbb{R}^n$; $\{v_1, v_2, \ldots, v_n\}$ are linearly independent and span V. Consider algorithm 1.6, where $v \cdot w$ denotes the dot-product of the two vectors. Show that algorithm 1.6 produces*

Algorithm 1.6 Gram-Schmidt

Pre-condition: $\{v_1, \ldots, v_n\}$ a basis for $\mathbb{R}^n$
1: $v_1^* \longleftarrow v_1$
2: **for** $i = 2, 3, \ldots, n$ **do**
3:      **for** $j = 1, 2, \ldots, (i - 1)$ **do**
4:          $\mu_{ij} \longleftarrow (v_i \cdot v_j^*)/\|v_j^*\|^2$
5:      **end for**
6:      $v_i^* \longleftarrow v_i - \sum_{j=1}^{i-1} \mu_{ij} v_j^*$
7: **end for**
Post-condition: $\{v_1^*, \ldots, v_n^*\}$ an orthogonal basis for $\mathbb{R}^n$

---

[2]A great and accessible introduction to linear algebra can be found in [Halmos (1995)].

an orthogonal basis $\{v_1^*, v_2^*, \ldots, v_n^*\}$ for the vector space V. In other words, show that $v_i^* \cdot v_j^* = 0$ when $i \neq j$, and that

$$span\{v_1, v_2, \ldots, v_n\} = span\{v_1^*, v_2^*, \ldots, v_n^*\}.$$

Justify why in line 4 of the algorithm we never divide by zero.

Based on the examples presented thus far it may appear that it is fairly clear to the naked eye whether an algorithm terminates or not, and the difficulty consists in coming up with a proof. But that is not the case.

Clearly, if we have a trivial algorithm consisting of a single while-loop, with the condition $i \geq 0$, and the body of the loop consists of the single command $i \longleftarrow i+1$, then we can immediately conclude that this while-loop will never terminate. But what about algorithm 1.7? Does it terminate?

Algorithm 1.7 Ulam's algorithm

Pre-condition: $a > 0$

$x \longleftarrow a$

while last three values of x not $4, 2, 1$ **do**

    if x is even **then**

        $x \longleftarrow x/2$

    else

        $x \longleftarrow 3x + 1$

    end if

end while

For example, if $a = 22$, then one can check that x takes on the following values: $22, 11, 34, 17, 52, 26, 13, 40, 20, 10, 5, 16, 8, \mathbf{4}, \mathbf{2}, \mathbf{1}$, and algorithm 1.7 terminates.

It is conjectured that regardless of the initial value of a, as long as a is a positive integer, algorithm 1.7 terminates. This conjecture is known as "Ulam's problem,"[3] No one has been able to prove that algorithm 1.7 terminates, and in fact proving termination would involve solving a difficult open mathematical problem.

---

[3]It is also called "Collatz Conjecture," "Syracuse Problem," "Kakutani's Problem," or "Hasse's Algorithm." While it is true that a rose by any other name would smell just as sweet, the preponderance of names shows that the conjecture is a very alluring mathematical problem.

1.3.5   *Recursion and fixed points*

So far we have proved the correctness of while-loops and for-loops, but there is another way of "looping" using *recursive* procedures, i.e., algorithms that "call themselves." We are going to see examples of such algorithms in the chapter on the divide and conquer method.

There is a robust theory of correctness of recursive algorithms based on fixed point theory, and in particular on Kleene's theorem (see appendix B, theorem B.31). We briefly illustrate this approach with an example. We are going to be using partial orders; all the necessary background can be found in appendix B, in section B.3. Consider the recursive algorithm 1.8.

Algorithm 1.8 $F(x, y)$

1: **if** $x = y$ **then**
2:         **return** $y + 1$
3: **else**
4:         $F(x, F(x - 1, y + 1))$
5: **end if**

To see how this algorithm works consider computing $F(4, 2)$. First in line 1 it is established that $4 \neq 2$ and so we must compute $F(4, F(3, 3))$. We first compute $F(3, 3)$, recursively, so in line 1 it is now established that $3 = 3$, and so in line 2 y is set to 4 and that is the value returned, i.e., $F(3, 3) = 4$, so now we can go back and compute $F(4, F(3, 3)) = F(4, 4)$, so again, recursively, we establish in line 1 that $4 = 4$, and so in line 2 y is set to 5 and this is the value returned, i.e., $F(4, 2) = 5$. On the other hand it is easy to see that

$$F(3, 5) = F(3, F(2, 6)) = F(3, F(2, F(1, 7))) = \cdots,$$

and this procedure never ends as x will never equal y. Thus F is not a *total* function, i.e., not defined on all $(x, y) \in \mathbb{Z} \times \mathbb{Z}$.

Problem 1.24. *What is the* domain of definition *of F as computed by algorithm 1.8? That is, the domain of F is $\mathbb{Z} \times \mathbb{Z}$, while the domain of definition is the largest subset $S \subseteq \mathbb{Z} \times \mathbb{Z}$ such that F is defined for all $(x, y) \in S$. We have seen already that $(4, 2) \in S$ while $(3, 5) \notin S$.*

We now consider three different functions, all given by algorithms that are not recursive: algorithms 1.9, 1.10 and 1.11, computing functions f_1, f_2 and f_3, respectively.

Algorithm 1.9 $f_1(x, y)$

---

if $x = y$ then
      return $y + 1$
else
      return $x + 1$
end if

Functions f_1 has an interesting property: if we were to replace F in algorithm 1.8 with f_1 we would get back F.

In other words, given algorithm 1.8, if we were to replace line 4 with $f_1(x, f_1(x-1, y+1))$, and compute f_1 with the (non-recursive) algorithm 1.9 for f_1, then algorithm 1.8 thus modified would now be computing $F(x, y)$. Therefore, we say that the functions f_1 is a *fixed point* of the recursive algorithm 1.8.

For example, recall the we have already shown that $F(4, 2) = 5$, using the recursive algorithm 1.8 for computing F. Replace line 4 in algorithm 1.8 with $f_1(x, f_1(x - 1, y + 1))$ and compute $F(4, 2)$ anew; since $4 \neq 2$ we go directly to line 4 where we compute $f_1(4, f_1(3, 3)) = f_1(4, 4) = 5$. Notice that this last computation was not recursive, as we computed f_1 directly with algorithm 1.9, and that we have obtained the same value.

Consider now f_2, f_3, computed by algorithms 1.10, 1.11, respectively.

Algorithm 1.10 $f_2(x, y)$

---

if $x \geq y$ then
      return $x + 1$
else
      return $y - 1$
end if

Algorithm 1.11 $f_3(x, y)$

---

if $x \geq y \wedge (x - y$ is even$)$ then
      return $x + 1$
end if

Notice that in algorithm 1.11, if it is not the case that $x \geq y$ and $x - y$ is even, then the output is undefined. Thus f_3 is a partial function, and if $x < y$ or $x - y$ is odd, then (x, y) is not in its domain of definition.

Problem 1.25. *Prove that f_1, f_2, f_3 are all fixed points of algorithm 1.8.*

The function f_3 has one additional property. For every pair of integers x, y such that $f_3(x, y)$ is defined, that is $x \geq y$ and $x - y$ is even, both $f_1(x, y)$ and $f_2(x, y)$ are also defined and have the same value as $f_3(x, y)$. We say that f_3 is *less defined than or equal to* f_1 and f_2, and write $f_3 \sqsubseteq f_1$ and $f_3 \sqsubseteq f_2$; that is, we have defined (informally) a partial order on functions $f : \mathbb{Z} \times \mathbb{Z} \longrightarrow \mathbb{Z} \times \mathbb{Z}$.

Problem 1.26. *Show that $f_3 \sqsubseteq f_1$ and $f_3 \sqsubseteq f_2$. Recall the notion of a domain of definition introduced in problem 1.24. Let S_1, S_2, S_3 be the domains of definition of f_1, f_2, f_3, respectively. You must show that $S_3 \subseteq S_1$ and $S_3 \subseteq S_2$.*

It can be shown that f_3 has this property, not only with respect to f_1 and f_2, but also with respect to all fixed points of algorithm 1.8. Moreover, $f_3(x, y)$ is the only function having this property, and therefore f_3 is said to be the *least (defined) fixed point of* algorithm 1.8. It is an important application of Kleene's theorem (theorem B.31) that every recursive algorithm has a unique fixed point. This unique fixed point may be seen as the "meaning" or "interpretation" of the recursive algorithm—and may be used for its proof of correctness.

This example illustrates an approach to the correctness of recursive algorithms; for more material we direct the interested reader to [Manna (1974)].

1.4   Stable marriage

We end this chapter with a very nice algorithm that is used in practice. It has two typical applications: the college admission process and matching interns with hospitals.

An instance of the *stable marriage problem* of size n consists of two disjoint finite sets of equal size; a set of *boys* $B = \{b_1, b_2, \ldots, b_n\}$, and a set of *girls* $G = \{g_1, g_2, \ldots, g_n\}$. Let "$<_i$" denote the ranking of boy b_i; that is, $g <_i g'$ means that boy b_i prefers g over g'. Similarly, "$<^j$" denotes the ranking of girl g_j. Each boy b_i has such a ranking (linear ordering) $<_i$ of G which reflects his preference for the girls that he wants to marry. Similarly each girl g_j has a ranking (linear ordering) $<^j$ of B which reflects her preference for the boys she would like to marry.

Fig. 1.3 Blocking pair.

A *matching* (or *marriage*) M is a 1-1 correspondence between B and G. We say that b and g are *partners* in M if they are matched in M and write $p_M(b) = g$ and also $p_M(g) = b$. A matching M is *unstable* if there is a pair (b, g) from $B \times G$ such that b and g are not partners in M but b prefers g to $p_M(b)$ and g prefers b to $p_M(g)$. Such a pair (b, g) is said to *block* the matching M and is called a *blocking pair* for M (see figure 1.3). A matching M is stable if there is no blocking pair for M.

A result of Gale and Shapley from 1962 is that any marriage problem has a solution, i.e., there always exists a stable matching, no matter what are the lists of preferences. In fact, they give an algorithm which produces a solution in n stages and takes $O(n^3)$ steps; we present their algorithm in this section (algorithm 1.12).

The matching M is produced in stages M_s so that b_t always has a partner at stage $s \geq t$ and $p_{M_t}(b_t) <_t p_{M_{t+1}}(b_t) <_t \cdots$. On the other hand, for each $g \in G$, if g has a partner at stage t, then g will have a partner at each stage $s \geq t$ and $p_{M_t}(g) >^t p_{M_{t+1}}(g) >^t \ldots$. Thus, as s increases, the partners of b_t become less preferable and the partners of g become more preferable.

At the end of stage s, assume that we have produced a matching

$$M_s = \{(b_1, g_{1,s}), \ldots, (b_s, g_{s,s})\},$$

where the notation $g_{i,s}$ means that $g_{i,s}$ is the partner of boy b_i after the end of stage s.

We will say that partners in M_s are *engaged*. The idea is that at stage $s+1$, b_{s+1} will try to get a partner by *proposing* to the girls in G in his order of preference. When b_{s+1} proposes to a girl g_j, g_j accepts his proposal if either g_j is not currently engaged or is currently engaged to a less preferable boy b, i.e., $b_{s+1} <^j b$. In the case where g_j prefers b_{s+1} over her current partner b, then g_j breaks off the engagement with b and b then has to search for a new partner.

Problem 1.27. *Show that each b need propose at most once to each g.*

Algorithm 1.12 Gale-Shapley

Stage 1: At stage 1, b_1 chooses the first girl g in his preference list and we set $M_1 = \{(b_1, g)\}$.

Stage $s + 1$:

$M \longleftarrow M_s$

$b^* \longleftarrow b_{s+1}$

Then b^* proposes to the girls in order of his preference until one accepts; girl g will accept the proposal as long as she is either not engaged or prefers b^* to her current partner $p_M(g)$.

Then we add (b^*, g) to M and proceed according to one of the following two cases:

(i) If g was not engaged, then we terminate the procedure and set $M_{s+1} \longleftarrow M \cup \{(b^*, g)\}$.

(ii) If g was engaged to b, then we set

$M \longleftarrow (M - \{(b, g)\}) \cup \{(b^*, g)\}$

$b^* \longleftarrow b$

and repeat.

From problem 1.27 we see that we can make each boy keep a bookmark on his list of preference, and this bookmark is only moving forward. When a boy's turn to choose comes, he starts proposing from the point where his bookmark is, and by the time he is done, his bookmark moved only forward. Note that at stage $s + 1$ each boy's bookmark cannot have moved beyond the girl number s on the list without choosing someone (after stage s only s girls are engaged). As the boys take turns, each boy's bookmark is advancing, so some boy's bookmark (among the boys in $\{b_1, \ldots, b_{s+1}\}$) will advance eventually to a point where he must choose a girl.

The discussion in the above paragraph shows that stage $s + 1$ in algorithm 1.12 must end. The concern here was that case (ii) of stage $s + 1$ might end up being circular. But the fact that the bookmarks are advancing shows that this is not possible.

Furthermore, this gives an upper bound of $(s+1)^2$ steps at stage $(s+1)$ in the procedure. This means that there are n stages, and each stage takes $O(n^2)$ steps, and hence algorithm 1.12 takes $O(n^3)$ steps altogether. The question, of course, is what do we mean by a step? Computers operate on binary strings, yet here the implicit assumption is that we compare numbers and access the lists of preferences in a single step. But the cost of these operations is negligible when compared to our idealized running time, and so we allow ourselves this poetic license to bound the overall running time.

Problem 1.28. *Show that there is exactly one girl that was not engaged at stage s but is engaged at stage $(s + 1)$ and that, for each girl g_j that is engaged in M_s, g_j will be engaged in M_{s+1} and that $p_{M_{s+1}}(g_j) <^j p_{M_s}(g_j)$. (Thus, once g_j becomes engaged, she will remain engaged and her partners will only gain in preference as the stages proceed.)*

Problem 1.29. *Suppose that $|B| = |G| = n$. Show that at the end of stage n, M_n will be a stable marriage.*

We say that a matching (b, g) is *feasible* if there exists a stable matching in which b, g are partners. We say that a matching is boy-*optimal* if every boy is paired with his highest ranked feasible partner. We say that a matching is boy-*pessimal* if every boy is paired with his lowest ranking feasible partner. Similarly, we define girl-*optimal/pessimal.*

Problem 1.30. *Show that our version of the algorithm produces a boy-optimal and girl-pessimal stable matching.*

1.5 Answers to selected problems

Problem 1.2. Basis case: $n = 1$, then $1^3 = 1^2$. For the induction step:

$$(1 + 2 + 3 + \cdots + n + (n + 1))^2$$
$$= (1 + 2 + 3 + \cdots + n)^2 + 2(1 + 2 + 3 + \cdots + n)(n + 1) + (n + 1)^2$$

and by the induction hypothesis,

$$= (1^3 + 2^3 + 3^3 + \cdots + n^3) + 2(1 + 2 + 3 + \cdots + n)(n + 1) + (n + 1)^2$$
$$= (1^3 + 2^3 + 3^3 + \cdots + n^3) + 2\frac{n(n + 1)}{2}(n + 1) + (n + 1)^2$$
$$= (1^3 + 2^3 + 3^3 + \cdots + n^3) + n(n + 1)^2 + (n + 1)^2$$
$$= (1^3 + 2^3 + 3^3 + \cdots + n^3) + (n + 1)^3$$

Problem 1.3. It is important to interpret the statement of the problem correctly; when it says that one square is missing, it means that any square can be missing. So the basis case is: given a 2×2 square, there are four possible ways for a square to be missing; but in each case, the remaining squares form an "L." These four possibilities are drawn in figure 1.4.

Fig. 1.4   The four different "L" shapes.

Suppose the claim holds for n, and consider a square of size $2^{n+1} \times 2^{n+1}$. Divide it into four quadrants of equal size. No matter which square we choose to be missing, it will be in one of the four quadrants; that quadrant can be filled with "L" shapes (i.e., shapes of the form given by figure 1.4) by induction hypothesis. As to the remaining three quadrants, put an "L" in them in such a way that it straddles all three of them (the "L" wraps around the center staying in those three quadrants). The remaining squares of each quadrant can now be filled with "L" shapes by induction hypothesis.

Problem 1.5. The basis case is $n = 1$, and it is immediate. For the induction step, assume the equality holds for exponent n, and show that it holds for exponent $n + 1$:

$$\begin{pmatrix} 1 & 1 \\ 1 & 0 \end{pmatrix}^n \begin{pmatrix} 1 & 1 \\ 1 & 0 \end{pmatrix} = \begin{pmatrix} f_{n+1} & f_n \\ f_n & f_{n-1} \end{pmatrix} \begin{pmatrix} 1 & 1 \\ 1 & 0 \end{pmatrix} = \begin{pmatrix} f_{n+1} + f_n & f_{n+1} \\ f_n + f_{n-1} & f_n \end{pmatrix}$$

The right-most matrix can be simplified using the definition of Fibonacci numbers to be as desired.

Problem 1.6. $m|n$ iff $n = km$, so show that $f_m | f_{km}$ by induction on k. If $k = 1$, there is nothing to prove. Otherwise, $f_{(k+1)m} = f_{km+m}$. Now, using a separate inductive argument, show that for $y \geq 1$, $f_{x+y} = f_y f_{x+1} + f_{y-1} f_x$, and finish the proof. To show this last statement, let $y = 1$, and note that $f_y f_{x+1} + f_{y-1} f_x = f_1 f_{x+1} + f_0 f_x = f_{x+1}$. Assume now that $f_{x+y} = f_y f_{x+1} + f_{y-1} f_x$ holds. Consider

$$\begin{aligned} f_{x+(y+1)} = f_{(x+y)+1} &= f_{(x+y)} + f_{(x+y)-1} = f_{(x+y)} + f_{x+(y-1)} \\ &= (f_y f_{x+1} + f_{y-1} f_x) + (f_{y-1} f_{x+1} + f_{y-2} f_x) \\ &= f_{x+1}(f_y + f_{y-1}) + f_x(f_{y-1} + f_{y-2}) \\ &= f_{x+1} f_{y+1} + f_x f_y. \end{aligned}$$

Problem 1.7. Note that this is almost the *Fundamental Theorem of Arithmetic*; what is missing is the fact that up to reordering of primes this representation is unique. The proof of this can be found in appendix A, theorem A.2.

Problem 1.8. Let our assertion $P(n)$ be: the minimal number of breaks to break up a chocolate bar of n squares is $(n - 1)$. Note that this says that $(n-1)$ breaks are sufficient, and $(n-2)$ are not. Basis case: only one square requires no breaks. Induction step: Suppose that we have $m+1$ squares. No matter how we break the bar into two smaller pieces of a and b squares each, $a+b=m+1$.

By induction hypothesis, the "a" piece requires $a-1$ breaks, and the "b" piece requires $b-1$ breaks, so together the number of breaks is

$$(a-1)+(b-1)+\boxed{1}=a+b-1=m-1,$$

and we are done (note that the 1 in the box comes from the initial break to divide the chocolate bar into the "a" and the "b" pieces).

So the "boring" way of breaking up the chocolate (first into rows, and then each row separately into pieces) is in fact optimal.

Problem 1.9. Let IP be: $[\mathrm{P}(0) \wedge (\forall n)(\mathrm{P}(n) \to \mathrm{P}(n+1))] \to (\forall m)\mathrm{P}(m)$ (where n, m range over natural numbers), and let LNP: *Every non-empty subset of the natural numbers has a least element.* These two principles are equivalent, in the sense that one can be shown from the other. Indeed:

LNP$\Rightarrow$IP: Suppose we have $[\mathrm{P}(0) \wedge (\forall n)(\mathrm{P}(n) \to \mathrm{P}(n+1))]$, but that it is *not* the case that $(\forall m)\mathrm{P}(m)$. Then, the set S of m's for which $\mathrm{P}(m)$ is false is non-empty. By the LNP we know that S has a least element. We know this element is not 0, as $\mathrm{P}(0)$ was assumed. So this element can be expressed as $n+1$ for some natural number n. But since $n+1$ is the least such number, $\mathrm{P}(n)$ must hold. This is a contradiction as we assumed that $(\forall n)(\mathrm{P}(n) \to \mathrm{P}(n+1))$, and here we have an n such that $\mathrm{P}(n)$ but not $\mathrm{P}(n+1)$.

IP$\Rightarrow$LNP: Suppose that S is a non-empty subset of the natural numbers. Suppose that it does not have a least element; let $\mathrm{P}(n)$ be the following assertion "all elements up to and including n are not in S." We know that $\mathrm{P}(0)$ must be true, for otherwise 0 would be in S, and it would then be the least element (by definition of 0). Suppose $\mathrm{P}(n)$ is true (so none of $\{0,1,2,\ldots,n\}$ is in S). Suppose $\mathrm{P}(n+1)$ were false: then $n+1$ would necessarily be in S (as we know that none of $\{0,1,2,\ldots,n\}$ is in S), and thereby $n+1$ would be the smallest element in S. So we have shown $[\mathrm{P}(0) \wedge (\forall n)(\mathrm{P}(n) \to \mathrm{P}(n+1))]$. By IP we can therefore conclude that $(\forall m)\mathrm{P}(m)$. But this means that S is empty. Contradiction. Thus S must have a least element.

Showing that IP and CIP are equivalent is straightforward.

Problem 1.10. We use the example in figure 1.1. Suppose that we want to obtain the tree from the infix (2164735) and prefix (1234675) encodings: from the prefix encoding we know that 1 is the root, and thus from the infix encoding we know that the left sub-tree has the infix encoding 2, and so prefix encoding 2, and the right sub-tree has the infix encoding 64735 and so prefix encoding 34675, and we proceed recursively.

Problem 1.11. Consider the following invariant: the sum S of the numbers currently in the set is odd. Now we prove that this invariant holds. Basis case: $S = 1+2+\cdots+2n = n(2n+1)$ which is odd. Induction step: assume S is odd, let S' be the result of one more iteration, so

$$S' = S + |a - b| - a - b = S - 2\min(a, b),$$

and since $2\min(a,b)$ is even, and S was odd by the induction hypothesis, it follows that S' must be odd as well. At the end, when there is just one number left, say x, $S = x$, so x is odd.

Problem 1.12. To solve this problem we must provide both an algorithm and an invariant for it. The algorithm works as follows: initially divide the club into any two groups. Let H be the total sum of enemies that each member has in his own group. Now repeat the following loop: while there is an m which has at least two enemies in his own group, move m to the other group (where m must have at most one enemy). Thus, when m switches houses, H decreases. Here the invariant is "H decreases monotonically." Now we know that a sequence of positive integers cannot decrease for ever, so when H reaches its absolute minimum, we obtain the required distribution.

Problem 1.13. At first, arrange the guests in any way; let H be the number of neighboring hostile pairs. We find an algorithm that reduces H whenever $H > 0$. Suppose $H > 0$, and let (A, B) be a hostile couple, sitting side-by-side, in the clockwise order A, B. Traverse the table, clockwise, until we find another couple (A', B') such that A, A' and B, B' are friends. Such a couple must exist: there are $2n - 2 - 1 = 2n - 3$ candidates for A' (these are all the people sitting clockwise after B, which have a neighbor sitting next to them, again clockwise, and that neighbor is neither A nor B). As A has at least n friends (among people other than itself), out of these $2n - 3$ candidates, at least $n - 1$ of them are friends of A. If each of these friends had an enemy of B sitting next to it (again, going clockwise), then B would have at least n enemies, which is not possible, so there must be an A' friends with A so that the neighbor of A' (clockwise) is B' and B' is a friend of B; see figure 1.5.

Note that when $n = 1$ no one has enemies, and so this analysis is applicable when $n \geq 2$, in which case $2n - 3 \geq 1$.

Now the situation around the table is $\ldots, A, \boxed{B, \ldots, A'}, B', \ldots$. Reverse everyone in the box (i.e., mirror image the box), to reduce H by 1. Keep

$$A, B, c_1, c_2, \ldots, c_{2n-3}, c_{2n-2}$$

Fig. 1.5   List of guests sitting around the table, in clockwise order, starting at A. We are interested in friends of A among $c_1, c_2, \ldots, c_{2n-3}$, to make sure that there is a neighbor to the right, and that neighbor is not A or B; of course, the table wraps around at c_{2n-2}, so the next neighbor, clockwise, of c_{2n-2} is A. As A has at most $n-1$ enemies, A has at least n friends (not counting itself; self-love does not count as friendship). Those n friends of A are among the c's, but if we exclude c_{2n-2} it follows that A has at least $n-1$ friends among $c_1, c_2, \ldots, c_{2n-3}$. If the clockwise neighbor of c_i, $1 \le i \le 2n-3$, i.e., c_{i+1} was in each case an enemy of B, then, as B already has an enemy of A, it would follow that B has n enemies, which is not possible.

repeating this procedure while $H > 0$; eventually $H = 0$ (by the LNP), at which point there are no neighbors that dislike each other.

Problem 1.14. We partition the participants into the set E of even persons and the set O of odd persons. We observe that, during the hand shaking ceremony, the set O cannot change its parity. Indeed, if two odd persons shake hands, O decreases by 2. If two even persons shake hands, O increases by 2, and, if an even and an odd person shake hands, $|O|$ does not change. Since, initially, $|O| = 0$, the parity of the set is preserved.

Problem 1.15. Suppose that $i|m$ and $i|n$. Then

$$i|(m - qn) = r = \text{rem}(m, n).$$

Therefore $i \le \gcd(n, \text{rem}(m, n))$, and as this is true for every i, it is in particular true for $i = \gcd(m, n)$; thus $\gcd(m, n) \le \gcd(n, \text{rem}(m, n))$. Conversely, suppose that $i|n$ and $i|\text{rem}(m, n)$. Then $i|m = qn + r$, so $i \le \gcd(m, n)$, and again, $\gcd(n, \text{rem}(m, n))$ meets the condition of being such an i, so we have $\gcd(n, \text{rem}(m, n)) \le \gcd(m, n)$. Both inequalities taken together give us $\gcd(m, n) = \gcd(n, \text{rem}(m, n))$.

Problem 1.16. Let r_i be r after the i-th iteration of the loop. Note that $r_0 = \text{rem}(m, n) = \text{rem}(a, b) \ge 0$, and in fact every $r_i \ge 0$ by definition of remainder. Furthermore:

$$r_{i+1} = \text{rem}(m', n') = \text{rem}(n, r) = \text{rem}(n, \text{rem}(m, n)) = \text{rem}(n, r_i) < r_i.$$

and so we have a decreasing, and yet non-negative, sequence of numbers; by the LNP this must terminate.

Problem 1.17. When $m < n$ then $\text{rem}(m, n) = m$, and so $m' = n$ and $n' = m$. Thus we execute one iteration of the loop only to swap m and n. Thus, we could add one line at the beginning of Euclid's algorithm to check if $m < n$, and if that is the case to swap them.

Problem 1.18. First we show that if $d = \gcd(a, b)$, then there exist u, v such that $au + bv = d$. Let $S = \{ax + by | ax + by > 0\}$; clearly $S \neq \emptyset$. By

LNP there exists a least $g \in S$. We show that $g = d$. Let $a = q \cdot g + r$, $0 \le r < g$. Suppose that $r > 0$; then

$$r = a - q \cdot g = a - q(ax_0 + by_0) = a(1 - qx_0) + b(-qy_0).$$

Thus, $r \in S$, but $r < g$—contradiction. So $r = 0$, and so $g|a$, and a similar argument shows that $g|b$. It remains to show that g is greater than any other common divisor of a, b. Suppose $c|a$ and $c|b$, so $c|(ax_0 + by_0)$, and so $c|g$, which means that $c \le g$. Thus $g = \gcd(a, b) = d$. For the extended Euclid algorithm itself see page 13, algorithm E, in [Knuth (1997)].

Problem 1.19. For partial correctness of algorithm 1.3, we show that if the pre-condition holds, and *if* the algorithm terminates, then the post-condition will hold. So assume the pre-condition, and suppose first that A is *not* a palindrome. Then there exists a smallest i_0 (there exists one, and so by the LNP there exists a smallest one) such that $A[i_0] \ne A[n - i_0 + 1]$, and so, after the first $i_0 - 1$ iteration of the while-loop, we know from the loop invariant that $i = (i_0 - 1) + 1 = i_0$, and so line 4 is executed and the algorithm returns F. Therefore, "A not a palindrome" $\Rightarrow$ "return F."

Suppose now that A *is* a palindrome. Then line 4 is never executed (as no such i_0 exists), and so after the $k = \lfloor \frac{n}{2} \rfloor$-th iteration of the while-loop, we know from the loop invariant that $i = \lfloor \frac{n}{2} \rfloor + 1$ and so the while-loop is not executed any more, and the algorithm moves on to line 8, and returns T. Therefore, "A is a palindrome" $\Rightarrow$ "return T."

Therefore, the post-condition, "return T iff A is a palindrome," holds. Note that we have only used part of the loop invariant, that is we used the fact that after the k-th iteration, $i = k + 1$; it still holds that after the k-th iteration, for $1 \le j \le k$, $A[j] = A[n - j + 1]$, but we do not need this fact in the above proof.

To show that the algorithm does actually terminates, let $d_i = \lfloor \frac{n}{2} \rfloor - i$. By the precondition, we know that $n \ge 1$. The sequence $d_1, d_2, d_3, \ldots$ is a decreasing sequence of positive integers (because $i \le \lfloor \frac{n}{2} \rfloor$), so by the LNP it is finite, and so the loop terminates.

Problem 1.20. The solution is given by algorithm 1.13.

Let the loop invariant be: "x is a power of 2 iff n is a power of 2."

We show the loop invariant by induction on the number of iterations of the main loop. Basis case: zero iterations, and since $x \leftarrow n$, $x = n$, so obviously x is a power of 2 iff n is a power of 2. For the induction step, note that if we ever get to update x, we have $x' = x/2$, and clearly x' is a power of 2 iff x is. Note that the algorithm always terminates (let $x_0 = n$, and $x_{i+1} = x_i/2$, and apply the LNP as usual).

Algorithm 1.13 Powers of 2.

Pre-condition: $n \geq 1$
  $x \leftarrow n$
  while $(x > 1)$ **do**
      if $(2|x)$ **then**
          $x \leftarrow x/2$
      else
          stop and return "no"
      end if
  end while
  return "yes"
Post-condition: "yes" $\iff$ n is a power of 2

We can now prove correctness: if the algorithms returns "yes", then after the last iteration of the loop $x = 1 = 2^0$, and by the loop invariant n is a power of 2. If, on the other hand, n is a power of 2, then so is every x, so eventually $x = 1$, and so the algorithm returns "yes".

Problem 1.21. Algorithm 1.4 computes the product of m and n, that is, the returned $z = m \cdot n$. A good loop invariant is $x \cdot y + z = m \cdot n$.

Problem 1.23. We are going to prove a loop invariant on the outer loop of algorithm 1.6, that is, we are going to prove a loop invariant on the for-loop (indexed on i) that starts on line 2 and ends on line 7. Our invariant consists of two parts: after the k-th iteration of the loop, the following two statements hold true:

(1) the set $\{v_1^*, \ldots, v_{k+1}^*\}$ is orthogonal, and
(2) span$\{v_1, \ldots, v_{k+1}\}$ = span$\{v_1^*, \ldots, v_{k+1}^*\}$.

Basis Case: after zero iterations of the for-loop, that is, before the for-loop is ever executed, we have, from line 1 of the algorithm, that $v_1^* \longleftarrow v_1$, and so the first statement is true because $\{v_1^*\}$ is orthogonal (a set consisting of a single non-zero vector is always orthogonal—and $v_1^* = v_1 \neq 0$ because the assumption (i.e., pre-condition) is that $\{v_1, \ldots, v_n\}$ is linearly independent, and so none of these vectors can be zero), and the second statement also holds trivially since if $v_1^* = v_1$ then span$\{v_1\}$ = span$\{v_1^*\}$.

Induction Step: Suppose that the two conditions hold after the first k iterations of the loop; we are going to show that they continue to hold after the $k+1$ iteration. Consider:

$$v_{k+2}^* = v_{k+2} - \sum_{j=1}^{k+1} \mu_{(k+1)j} v_j^*,$$

which we obtain directly from line 6 of the algorithm; note that the outer for-loop is indexed on i which goes from 2 to n, so after the k-th execution of line 2, for $k \geq 1$, the value of the index i is $k + 1$. We show the first statement, i.e., that $\{v_1^*, \ldots, v_{k+2}^*\}$ are orthogonal. Since, by induction hypothesis, we know that $\{v_1^*, \ldots, v_{k+1}^*\}$ are already orthogonal, it is enough to show that for $1 \leq l \leq k + 1$, $v_l^* \cdot v_{k+2}^* = 0$, which we do next:

$$v_l^* \cdot v_{k+2}^* = v_l^* \cdot \left(v_{k+2} - \sum_{j=1}^{k+1} \mu_{(k+2)j} v_j^* \right)$$

$$= (v_l^* \cdot v_{k+2}) - \sum_{j=1}^{k+1} \mu_{(k+2)j} (v_l^* \cdot v_j^*)$$

and since $v_l^* \cdot v_j^* = 0$ unless $l = j$, we have:

$$= (v_l^* \cdot v_{k+2}) - \mu_{(k+2)l} (v_l^* \cdot v_l^*)$$

and using line 4 of the algorithm we write:

$$= (v_l^* \cdot v_{k+2}) - \frac{v_{k+2} \cdot v_l^*}{\|v_l^*\|^2} (v_l^* \cdot v_l^*) = 0$$

where we have used the fact that $v_l \cdot v_l = \|v_l\|^2$ and that $v_l^* \cdot v_{k+2} = v_{k+2} \cdot v_l^*$.

For the second statement of the loop invariant we need to show that

$$\text{span}\{v_1, \ldots, v_{k+2}\} = \text{span}\{v_1^*, \ldots, v_{k+2}^*\}, \tag{1.5}$$

assuming, by the induction hypothesis, that $\text{span}\{v_1, \ldots, v_{k+1}\} = \text{span}\{v_1^*, \ldots, v_{k+1}^*\}$. The argument will be based on line 6 of the algorithm, which provides us with the following equality:

$$v_{k+2}^* = v_{k+2} - \sum_{j=1}^{k+1} \mu_{(k+2)j} v_j^*. \tag{1.6}$$

Given the induction hypothesis, to show (1.5) we need only show the following two things:

(1) $v_{k+2} \in \text{span}\{v_1^*, \ldots, v_{k+2}^*\}$, and
(2) $v_{k+2}^* \in \text{span}\{v_1, \ldots, v_{k+2}\}$.

Using (1.6) we obtain immediately that $v_{k+2} = v^*_{k+2} + \sum_{j=1}^{k+1} \mu_{(k+2)j} v^*_j$ and so we have (1). To show (2) we note that

$$\text{span}\{v_1, \ldots, v_{k+2}\} = \text{span}\{v^*_1, \ldots, v^*_{k+1}, v_{k+2}\}$$

by the induction hypothesis, and so we have what we need directly from (1.6).

Finally, note that we never divide by zero in line 4 of the algorithm because we always divide by $\|v^*_j\|$, and the only way for the norm to be zero is if the vector itself, v^*_j, is zero. But we know from the post-condition that $\{v^*_1, \ldots, v^*_n\}$ is a basis, and so these vectors must be linearly independent, and so none of them can be zero.

Problem 1.24. Let $S \subseteq \mathbb{Z} \times \mathbb{Z}$ be the set consisting precisely of those pairs of integers (x, y) such that $x \geq y$ and $x - y$ is even. We are going to prove that S is the domain of definition of F. First, if $x < y$ then $x \neq y$ and so we go on to compute $F(x, F(x - 1, y + 1))$, and now we must compute $F(x - 1, y + 1)$; but if $x < y$, then clearly $x - 1 < y + 1$; this condition is preserved, and so we end up having to compute $F(x - i, y + i)$ for all i, and so this recursion never "bottoms out." Suppose that $x - y$ is odd. Then $x \neq y$ (as 0 is even!), so again we go on to $F(x, F(x - 1, y + 1))$; if $x - y$ is odd, so is $(x - 1) - (y + 1) = x - y - 2$. Again we end up having to compute $F(x - i, y + i)$ for all i, and so the recursion never terminates. Clearly, all the pairs in S^c are not in the domain of definition of F.

Suppose now that $(x, y) \in S$. Then $x \geq y$ and $x - y$ is even; thus, $x - y = 2i$ for some $i \geq 0$. We show, by induction on i, that the algorithm terminates on such (x, y) and outputs $x + 1$. Basis case: $i = 0$, so $x = y$, and so the algorithm returns $y + 1$ which is $x + 1$. Suppose now that $x - y = 2(i + 1)$. Then $x \neq y$, and so we compute $F(x, F(x - 1, y + 1))$. But

$$(x - 1) - (y + 1) = x - y - 2 = 2(i + 1) - 2 = 2i,$$

for $i \geq 0$, and so by induction $F(x - 1, y + 1)$ terminates and outputs $(x - 1) + 1 = x$. So now we must compute $F(x, x)$ which is just $x + 1$, and we are done.

Problem 1.25. We show that f_1 is a fixed point of algorithm 1.8. Recall that in problem 1.24 we showed that the domain of definition of F, the function computed by algorithm 1.8, is $S = \{(x, y) : x - y = 2i, i \geq 0\}$. Now we show that if we replace F in algorithm 1.8 by f_1, the new algorithm, which is algorithm 1.14, still computes F albeit not recursively (as f_1 is defined by algorithm 1.9 which is not recursive).

Algorithm 1.14 Algorithm 1.8 with F replaced by f_1.

1: **if** $x = y$ **then**
2:         **return** $y + 1$
3: **else**
4:         $f_1(x, f_1(x - 1, y + 1))$
5: **end if**

We proceed as follows: if $(x, y) \in S$, then $x - y = 2i$ with $i \geq 0$. On such (x, y) we know, from problem 1.24, that $F(x, y) = x + 1$. Now consider the output of algorithm 1.14 on such a pair (x, y). If $i = 0$, then it returns $y + 1 = x + 1$, so we are done. If $i > 0$, then it computes

$$f_1(x, f_1(x - 1, y + 1)) = f_1(x, x) = x + 1,$$

and we are done. To see why $f_1(x - 1, y + 1) = x$ notice that there are two cases: first, if $x - 1 = y + 1$, then the algorithm for f_1 (algorithm 1.9) returns $(y + 1) + 1 = (x - 1) + 1 = x$. Second, if $x - 1 > y + 1$ (and that is the only other possibility), algorithm 1.9 returns $(x - 1) + 1 = x$ as well.

Problem 1.27. After b proposed to g for the first time, whether this proposal was successful or not, the partners of g could have only gotten better. Thus, there is no need for b to try again.

Problem 1.28. b_{s+1} proposes to the girls according to his list of preference; a g ends up accepting, and if the g who accepted b_{s+1} was free, she is the new one with a partner. Otherwise, some $b^* \in \{b_1, \ldots, b_s\}$ became disengaged, and we repeat the same argument. The g's disengage only if a better b proposes, so it is true that $p_{M_{s+1}}(g_j) <^j p_{M_s}(g_j)$.

Problem 1.29. Suppose that we have a blocking pair $\{b, g\}$ (meaning that $\{(b, g'), (b', g)\} \subseteq M_n$, but b prefers g to g', and g prefers b to b'). Either b came after b' or before. If b came before b', then g would have been with b or someone better when b' came around, so g would not have become engaged to b'. On the other hand, since (b', g) is a pair, no better offer has been made to g after the offer of b', so b could not have come after b'. In either case we get an impossibility, and so there is no blocking pair $\{b, g\}$.

Problem 1.30. To show that the matching is boy-optimal, we argue by contradiction. Let "*g is an optimal partner for b*" mean that among all the stable matchings g is the best partner that b can get.

We run the Gale-Shapley algorithm, and let b be the first boy who is rejected by his optimal partner g. This means that g has already been paired with some b', and g prefers b' to b. Furthermore, g is at least as desirable to b' as his own optimal partner (since the proposal of b is the first time during the run of the algorithm that a boy is rejected by his optimal partner). Since g is optimal for b, we know (by definition) that there exists some stable matching S where (b, g) is a pair. On the other hand, the optimal partner of b' is ranked (by b' of course) at most as high as g, and since g is taken by b, whoever b' is paired with in S, say g', b' prefers g to g'. This gives us an unstable pairing, because $\{b', g\}$ prefer each other to the partners they have in S.

To show that the Gale-Shapley algorithm is girl-pessimal, we use the fact that it is boy-optimal (which we just showed). Again, we argue by contradiction. Suppose there is a stable matching S where g is paired with b, and g prefers b' to b, where (b', g) is the result of the Gale-Shapley algorithm. By boy-optimality, we know that in S we have (b', g'), where g' is not higher on the preference list of b' than g, and since g is already paired with b, we know that g' is actually lower. This says that S is unstable since $\{b', g\}$ would rather be together than with their partners.

1.6 Notes

This book is about proving things about algorithms; their correctness, their termination, their running time, etc. The art of mathematical proofs is a difficult art to master; a very good place to start is [Velleman (2006)].

$\mathbb{N}$ (the set of natural numbers) and IP (the induction principle) are very tightly related; the rigorous definition of $\mathbb{N}$, as a set-theoretic object, is the following: it is the *unique* set satisfying the following three properties: (i) it contains 0, (ii) if n is in it, then so is $n+1$, and (iii) it satisfies the induction principle (which in this context is stated as follows: if S is a subset of $\mathbb{N}$, and S satisfies (i) and (ii) above, then in fact $S = \mathbb{N}$).

The references in this paragraph are with respect to Knuth's seminal *The Art of Computer Programming*, [Knuth (1997)]. For an extensive study of Euclid's algorithm see §1.1. Problem 1.2 comes from §1.2.1, problem #8, pg. 19. See §2.3.1, pg. 318 for more background on tree traversals. For the history of the concept of pre and post-condition, and loop invariants, see pg. 17.

See [Zingaro (2008)] for a book dedicated to the idea of invariants in the context of proving correctness of algorithms. A great source of problems on the invariance principle, that is section 1.2, is chapter 1 in [Engel (1998)]

The example about the 8×8 board with two squares missing (figure 1.2) comes from [Dijkstra (1989)].

The palindrome `madamimadam` comes from Joyce's *Ulysses*.

Section 1.3.5 on the correctness of recursive algorithms is based on chapter 5 of [Manna (1974)].

Section 1.4 is based on §2 in [Cenzer and Remmel (2001)]. For another presentation of the Stable Marriage problem see chapter 1 in [Kleinberg and Tardos (2006)].

Chapter 2

Greedy Algorithms

Greedy algorithms are algorithms prone to instant gratification. Without looking too far ahead, at each step they make a *locally optimum* choice, with the hope that it will lead to a *global optimum* at the end.

An example of a greedy procedure is a convenience store clerk dispensing change. In order to use as few coins as possible, the clerk gives out the coins of highest value for as long as possible, before moving on to a smaller denomination.

Being greedy is a simple heuristic that works well with some computational problems but fails with others. In the case of cash dispensing, if we have coins of value $1, 5, 25$ the greedy procedure always produces the smallest possible (i.e., optimal) number of coins. But if we have coins of value $1, 10, 25$, then dispensing 30 greedily results in six coins $(25, 1, 1, 1, 1, 1)$, while a more astute—and less greedy—clerk would give out only three coins $(10, 10, 10)$. See section 2.3.1 for the make-change problem.

2.1 Minimum cost spanning trees

We start by giving several graph-theoretic definitions. Note that our graphs are assumed to be finite (i.e., they have finitely many nodes).

An *undirected graph* G is a pair (V, E) where V is a set (of vertices or nodes), and $E \subseteq V \times V$ and $(u, v) \in E$ iff $(v, u) \in E$, and $(u, u) \notin E$. The *degree* of a vertex v is the number of edges touching v. A *path* in G between v_1 and v_k is a sequence $v_1, v_2, \ldots, v_k$ such that each $(v_i, v_{i+1}) \in E$. G is *connected* if between every pair of distinct nodes there is a path. A *cycle* is a simply closed path $v_1, \ldots, v_k, v_1$ with $v_1, \ldots, v_k$ all distinct, and $k \geq 3$. A graph is *acyclic* if it has no cycles. A *tree* is a connected acyclic graph. A *spanning tree* of a connected graph G is a subset $T \subseteq E$ of the edges such that (V, T) is a tree. In other words, the edges in T must connect all nodes of G and contain no cycles.

If G has a cycle, then there is more than one spanning tree for G, and in general G may have many spanning trees, but each spanning tree has the same number of edges.

Lemma 2.1. *Every tree with n nodes has exactly $n - 1$ edges.*

Problem 2.2. *Prove lemma 2.1. (Hint: first show that every tree has a leaf, i.e., a node of degree one. Then show the lemma by induction on n.)*

Lemma 2.3. *A graph with n nodes and more than $n - 1$ edges must contain at least one cycle.*

Problem 2.4. *Prove lemma 2.3.*

It follows from lemmas 2.1 and 2.3 that if a graph is a tree, i.e., it is acyclic and connected, then it must have $(n - 1)$ edges. If it does *not* have $(n - 1)$ edges, then it is either not acyclic, or it is not connected. If it has less than $(n - 1)$ edges, it is certainly not connected, and if it has more than $(n - 1)$ edges, it is certainly not acyclic.

We are interested in finding a minimum cost spanning tree for G, assuming that each edge e is assigned a cost $c(e)$. The understanding is that the costs are non-negative real number, i.e., each $c(e)$ is in $\mathbb{R}^+$. The total cost $c(T)$ is the sum of the costs of the edges in T. We say that T is a *minimum cost spanning tree (MCST)* for G if T is a spanning tree for G and given any spanning tree T' for G, $c(T) \leq c(T')$.

Given a graph $G = (V, E)$, where $c(e_i) = $ "cost of edge e_i," we want to find a MCST. It turns out (fortunately) that in this case, an obvious greedy algorithm (Kruskal's algorithm) always works. Kruskal's algorithm is the following: sort the edges in non-decreasing order of costs, so that $c(e_1) \leq c(e_2) \leq \ldots \leq c(e_m)$, add the edges one at a time, except when adding an edge would form a cycle.

But how do we test for a cycle (i.e., execute line 4)? At the end of each iteration of the for-loop, the set T of edges divides the vertices V into a collection $V_1, \ldots, V_k$ of *connected components*. That is, V is the disjoint union of $V_1, \ldots, V_k$, each V_i forms a connected graph using edges from T, and no edge in T connects V_i and V_j, if $i \neq j$. A simple way to keep track of $V_1, \ldots, V_k$ is to use an array $D[i]$ where $D[i] = j$ if vertex $i \in V_j$. Initialize D by setting $D[i] \longleftarrow i$ for every $i = 1, 2, \ldots, n$.

Algorithm 2.1 Kruskal

1: Sort the edges: $c(e_1) \le c(e_2) \le \ldots \le c(e_m)$
2: $T \longleftarrow \emptyset$
3: **for** $i : 1..m$ **do**
4:     **if** $T \cup \{e_i\}$ has no cycle **then**
5:         $T \longleftarrow T \cup \{e_i\}$
6:     **end if**
7: **end for**

To check whether $e_i = (r, s)$ forms a cycle within T, it is enough to check whether $D[r] = D[s]$. If e_i does not form a cycle within T, then we update: $T \longleftarrow T \cup \{(r, s)\}$, and we merge the component $D[r]$ with $D[s]$ as shown in algorithm 2.2.

Algorithm 2.2 Merging components

$k \longleftarrow D[r]$
$l \longleftarrow D[s]$
for $j : 1..n$ **do**
    if $D[j] = l$ **then**
        $D[j] \longleftarrow k$
    end if
end for

Problem 2.5. *Given that the edges can be ordered in m^2 steps (using insertion sort, or in $O(m \log m)$ using something more fancy like heapsort, and assuming again that the cost of a comparison is that of a single operation), what is the running time of algorithm 2.1?*

It is not immediately clear that Kruskal's algorithm yields a spanning, let alone a MCST. To see that the resulting collection T of edges is a spanning tree for G (assuming that G is connected), we must show that (V, T) is connected and acyclic.

It is obvious that T is acyclic, because we never add an edge that results in a cycle. To show that (V, T) is connected, we reason as follows. Let u and v be two distinct nodes in V. Since G is connected, there is a path p connecting u and v in G. The algorithm considers each edge e_i of G in turn, and puts e_i in T *unless* $T \cup \{e_i\}$ forms a cycle. But in the latter case, there must already be a path in T connecting the end points of e_i, so deleting e_i

does not disconnect the graph.

This argument can be formalized by showing that the following statement is an invariant of the loop in Kruskal's algorithm:

$$\text{The edge set } T \cup \{e_{i+1}, \ldots, e_m\} \text{ connects all nodes in } V. \qquad (2.1)$$

Problem 2.6. *Prove, by induction, that (2.1) is a loop invariant. First show that (2.1) holds before the main loop of Kruskal's algorithm executes (the 0-th iteration; this is the basis case—remember the assumption (precondition) that $G = (V, E)$ is connected). Then show that if (2.1) holds after the i-th execution of the loop, then $T \cup \{e_{i+2}, \ldots, e_m\}$ connects all nodes of V after the $(i+1)$-st execution of the loop. Conclude by induction that (2.1) holds for all i. Finally, show how to use this loop invariant to prove that T is connected.*

Problem 2.7. *Suppose that $G = (V, E)$ is not connected. Show that in this case, when G is given to Kruskal's algorithm as input, the algorithm computes a spanning forest of G. Define first precisely what is a spanning forest (first define the notion of a connected component). Then give a formal proof using the idea of a loop invariant, as in problem 2.6.*

To show that the spanning tree resulting from the algorithm is in fact a MCST, we reason that after each iteration of the loop, the set T of edges can be extended to a MCST using edges that have *not yet* been considered. Hence after termination, all edges have been considered, so T must itself be a MCST. We say that a set T of edges of G is *promising* if T can be extended to a MCST for G, that is, T is promising if there exists a MCST T' such that $T \subseteq T'$.

Lemma 2.8. *"T is promising" is a loop invariant for Kruskal's algorithm.*

Proof.   The proof is by induction on the number of iterations of the main loop of Kruskal's algorithm. Basis case: at this stage the algorithm has gone through the loop zero times, and initially T is the empty set, which is obviously promising (the empty set is a subset of any set).

Induction step: We assume that T is promising, and show that T continues being promising after one more iteration of the loop.

Notice that the edges used to expand T to a spanning tree must come from those not yet considered, because the edges that have been considered are either in T already, or have been rejected because they form a cycle. We examine by cases what happens after edge e_i has been considered:

Case 1: e_i is rejected. T remains unchanged, and it is still promising. There is one subtle point: T was promising before the loop was executed, meaning that there was a subset of edges $S \subseteq \{e_i, \ldots, e_m\}$ that extended T to a MCST, i.e., $T \cup S$ is a MCST. But after the loop is executed, the edges extending T to a MCST would come from $\{e_{i+1}, \ldots, e_m\}$; but this is not a problem, as e_i could not be part of S (as then $T \cup S$ would contain a cycle), so $S \subseteq \{e_{i+1}, \ldots, e_m\}$, and so S is still a candidate for extending T to a MCST, even *after* the execution of the loop. Thus T remains promising after the execution of the loop, though the edges extending it to a MCST come from a smaller set (i.e., not containing e_i).

Case 2: e_i is accepted. We must show that $T \cup \{e_i\}$ is still promising. Since T is promising, there is a MCST T_1 such that $T \subseteq T_1$. We consider two subcases.

Subcase a: $e_i \in T_1$. Then obviously $T \cup \{e_i\}$ is promising.

Subcase b: $e_i \notin T_1$. Then, according to the Exchange Lemma below, there is an edge e_j in $T_1 - T_2$ (where T_2 is the spanning tree resulting from the algorithm) such that $T_3 = (T_1 \cup \{e_i\}) - \{e_j\}$ is a spanning tree. Notice that $i < j$, since otherwise e_j would have been rejected from T and thus would form a cycle in T and so also in T_1. Therefore $c(e_i) \leq c(e_j)$, so $c(T_3) \leq c(T_1)$, so T_3 must also be a MCST. Since $T \cup \{e_i\} \subseteq T_3$, it follows that $T \cup \{e_i\}$ is promising.

This finishes the proof of the induction step. ☐

Lemma 2.9 (Exchange Lemma). *Let G be a connected graph, and let T_1 and T_2 be any two spanning trees for G. For every edge e in $T_2 - T_1$ there is an edge e' in $T_1 - T_2$ such that $T_1 \cup \{e\} - \{e'\}$ is a spanning tree for G.*

Problem 2.10. *Prove this lemma. (Hint: let e be an edge in $T_2 - T_1$. Then $T_1 \cup \{e\}$ contains a cycle—can all the edges in this cycle belong to T_2?).*

Problem 2.11. *Suppose that edge e_1 has a smaller cost than any of the other edges; that is, $c(e_1) < c(e_i)$, for all $i > 1$. Show that there is at least one MCST for G that includes edge e_1.*

Problem 2.12. *Kruskal's algorithm can return different spanning trees for the same input graph G, depending on how ties are broken when the edges are sorted (remember that before the algorithm 2.1 proceeds, it orders the edges in line 1, and presumably breaks ties—i.e., sorts edges of the same cost—arbitrarily). Show that for every MCST T of a graph G, there exists* *a way of sorting the edges (i.e., breaking the ties) so that the algorithm returns T.*

2.2   Jobs with deadlines and profits

We have n jobs, each of which takes unit time, and a processor on which we would like to schedule them in as profitable a manner as possible. Each job has a profit associated with it, as well as a deadline; if the job is not scheduled by its deadline, then we do not get its profit. Because each job takes the same amount of time, we think of a schedule S as consisting of a sequence of job "slots" $1, 2, 3, \ldots$, where $S(t)$ is the job scheduled in slot t.

Formally, the input is a sequence of pairs $(d_1, g_1), (d_2, g_2), \ldots, (d_n, g_n)$ where $g_i \in \mathbb{R}^+$ is the profit obtainable from job i, and $d_i \in \mathbb{N}$ is the deadline for job i.

A *schedule* is an array $S(1), S(2), \ldots, S(d)$ where $d = \max d_i$, that is, d is the latest deadline, beyond which no jobs can be scheduled. If $S(t) = i$, then job i is scheduled at time t, $1 \le t \le d$. If $S(t) = 0$, then no job is scheduled at time t. A schedule S is *feasible* if it satisfies two conditions: **(a)** if $S(t) = i > 0$, then $t \le d_i$ (i.e., every scheduled job meets its deadline), and **(b)** if $t_1 \ne t_2$ and $S(t_1) \ne 0$, then $S(t_1) \ne S(t_2)$ (i.e., each job is scheduled at most once). Let the total profit of schedule S be $P(S) = \sum_{t=1}^{d} g_{S(t)}$, where $g_0 = 0$.

We want to find a feasible schedule S whose profit $P(S)$ is as large as possible. This can be accomplished with the greedy algorithm 2.3.

Algorithm 2.3 Job scheduling

1: Sort the jobs in non-increasing order of profits: $g_1 \ge g_2 \ge \cdots \ge g_n$
2: $d \longleftarrow \max_i d_i$
3: **for** $t : 1..d$ **do**
4:       $S(t) \longleftarrow 0$
5: **end for**
6: **for** $i : 1..n$ **do**
7:       Find the largest t such that $S(t) = 0$ and $t \le d_i$, $S(t) \longleftarrow i$
8: **end for**

Note that line 7 finds the latest possible free slot meeting the deadline; if no such free slot exists, then job i cannot be scheduled. More precisely, the understanding is that if there is no t satisfying $S(t) = 0$ and $t \le d_i$, then the last command on line 7, $S(t) \longleftarrow i$, is *not* executed, and the for-loop considers the next i.

Theorem 2.13. *The greedy solution to job scheduling is optimal, that is, the profit $P(S)$ of the schedule S computed by algorithm 2.3 is as large as possible.*

A schedule is *promising* if it can be extended to an optimal schedule. Schedule S' *extends* schedule S iff for all $1 \le t \le d$, if $S(t) \ne 0$, then $S(t) = S'(t)$. For example, if $S = (2, 0, 0, 0, 3)$ and $S' = (2, 0, 1, 0, 3)$, then S' extends S.

Lemma 2.14. *"S is promising" is an invariant for the (second) for-loop in algorithm 2.3.*

In fact, we must make the definition of "promising" in lemma 2.14 more precise: we say that "S is promising *after* the i-th iteration of the loop in algorithm 2.3" if S can be extended to an optimal schedule using jobs from those among $\{i + 1, i + 2, \ldots, n\}$, i.e., using only those jobs that have not been considered yet by algorithm 2.3.

Problem 2.15. *Why does lemma 2.14 imply theorem 2.13? (Hint: this is a simple observation).*

We now prove lemma 2.14.

Proof. The proof is by induction. Basis case: after the 0-th iteration of the loop, $S = (0, 0, \ldots, 0)$ and we may extend it with jobs $\{1, 2, \ldots, n\}$, i.e., we have all the jobs at our disposal; so S is promising, as we can take *any* optimal schedule, and it will an extension of S.

Induction step: Suppose that S is promising, and let S_{opt} be *some* optimal schedule that extends S. Let S' be the result of one more iteration through the loop where job i is considered. We must prove that S' continues being promising, so the goal is to show that there is an optimal schedule S'_{opt} that extends S'. We consider two cases:

Case 1: job i cannot be scheduled. Then $S' = S$, so we let $S'_{\text{opt}} = S_{\text{opt}}$, and we are done. The only subtle thing is that S was extendable into S_{opt} with jobs in $\{i, i + 1, \ldots, n\}$, but after the i-th iteration we no longer have job i at our disposal.

Problem 2.16. *Show that this "subtle thing" mentioned in the paragraph above is not a problem.*

Case 2: job i is scheduled at time t_0, so $S'(t_0) = i$ (whereas $S(t_0) = 0$) and t_0 is the latest possible time for job i in the schedule S. We have two subcases.

Subcase a: job i is scheduled in S_{opt} at time t_1:

If $t_1 = t_0$, then, as in case 1, just let $S'_{\text{opt}} = S_{\text{opt}}$.

If $t_1 < t_0$, then let S'_{opt} be S_{opt} except that we interchange t_0 and t_1, that is we let $S'_{\text{opt}}(t_0) = S_{\text{opt}}(t_1) = i$ and $S'_{\text{opt}}(t_1) = S_{\text{opt}}(t_0)$. Then S'_{opt} is feasible (why 1?), it extends S' (why 2?), and $P(S'_{\text{opt}}) = P(S_{\text{opt}})$ (why 3?).

The case $t_1 > t_0$ is not possible (why 4?).

Subcase b: job i is not scheduled in S_{opt}. Then we simply define S'_{opt} to be the same as S_{opt}, except $S'_{\text{opt}}(t_0) = i$. Since S_{opt} is feasible, so is S'_{opt}, and since S'_{opt} extends S', we only have to show that $P(S'_{\text{opt}}) = P(S_{\text{opt}})$. This follows from the following claim:

Claim 2.17. *Let* $S_{\text{opt}}(t_0) = j$. *Then* $g_j \leq g_i$.

Proof. We prove the claim by contradiction: assume that $g_j > g_i$ (note that in this case $j \neq 0$). Then job j was considered before job i. Since job i was scheduled at time t_0, job j must have been scheduled at time $t_2 \neq t_0$ (we know that job j was scheduled in S since $S(t_0) = 0$, and $t_0 \leq d_j$, so there was a slot for job j, and therefore it was scheduled). But S_{opt} extends S, and $S(t_2) = j \neq S_{\text{opt}}(t_2)$—contradiction. □

This finishes the proof of the induction step. □

Problem 2.18. *Make sure you can answer all the "why's" in the above proof. Also, where in the proof of the claim you use the fact that* $j \neq 0$?

## 2.3	Further examples and problems

### 2.3.1	*Make change*

The make-change problem, briefly described in the introduction to this chapter, consists in paying a given amount using the least number of coins, using some fixed denomination, and an unlimited supply of coins of each denomination.

Consider the following greedy algorithm to solve the make-change problem, where the denominations are $C = \{1, 10, 25, 100\}$. On input $n \in \mathbb{N}$, the algorithm outputs the smallest list L of coins (from among C) whose sum equals n.

Algorithm 2.4 Make change

1: $C \longleftarrow \{1, 10, 25, 100\}$; $L \longleftarrow \emptyset$; $s \longleftarrow 0$
2: **while** $(s < n)$ **do**
3:       find the largest x in C such that $s + x \leq n$
4:       $L \longleftarrow L \cup \{x\}$; $s \longleftarrow s + x$
5: **end while**
6: **return** L

Note that s equals the sum of the values of the coins in L, and that strictly speaking L is a *multiset* (the same element may appear more than once in a multiset).

Problem 2.19. *Show that algorithm 2.4 (with the given denominations) does not necessarily produce an optimal solution. That is, present an n for which the output L contains more coins than the optimal solution.*

Problem 2.20. *Suppose that $C = \{1, p, p^2, \ldots, p^n\}$, where $p > 1$ and $n \geq 0$ are integers. That is, "$C \longleftarrow \{1, 10, 25, 100\}$" in line 1 of algorithm 2.4 is replaced by "$C \longleftarrow \{1, p, p^2, \ldots, p^n\}$." Show that with this series of denominations (for some fixed p, n) the greedy algorithm above always finds an optimal solution. (Hint: Start with a suitable definition of a promising list.)*

2.3.2 *Maximum weight matching*

Let $G = (V_1 \cup V_2, E)$ be a bipartite graph, with edge set $E \subseteq V_1 \times V_2$, and $w : E \longrightarrow \mathbb{N}$ assigns a weight $w(e) \in \mathbb{N}$ to each edge $e \in E = \{e_1, \ldots, e_m\}$. A *matching* for G is a subset $M \subseteq E$ such that no two edges in M share a common vertex. The weight of M is $w(M) = \sum_{e \in M} w(e)$.

Problem 2.21. *Give a simple greedy algorithm which, given a bipartite graph with edge weights, attempts to find a matching with the largest possible weight.*

Problem 2.22. *Give an example of a bipartite graph with edge weights for which your algorithm in problem 2.21 fails to find a matching with the largest possible weight.*

Problem 2.23. *Suppose all edge weights in the bipartite graph are distinct, and each is a power of 2. Prove that your greedy algorithm always* *succeeds in finding a maximum weight matching in this case. (Assume for this question that all the edges are there, i.e., that $E = V \times V$.)*

2.3.3 Shortest path

Suppose that we are given a graph $G = (V, E)$, a designated start node s, and a cost function for each edge $e \in E$, denoted $c(e)$. We are asked to compute the cheapest paths from s to every other node in G, where the cost of a path is the sum of the costs of its edges.

Consider the following greedy algorithm: the algorithm maintains a set S of explored nodes, and for each $u \in S$ it stores a value $d(u)$, which is the cheapest path distance from s to u.

Initially, $S = \{s\}$ and $d(s) = 0$. Now, for each $v \in V - S$ we find the shortest path that can be constructed by traveling along a path through the explored part of S to some $u \in S$, followed by a single edge (u, v). That is, we consider the quantity $d'(v) = \min_{e=(u,v), u \in S} d(u) + c(e)$. We choose the node $v \in V - S$ for which this quantity is minimized, add v to S, and set $d(v) = d'(v)$, and repeat, and stop when $S = V$. See figure 2.1.

Fig. 2.1 Computing the shortest path.

This greedy algorithm for computing the shortest path is due to Edsger Dijkstra. It is not difficult to see that its running time is $O(n^2)$.

Problem 2.24. *Show that at the end, for each $u \in V$, $d(u)$ is the cost of the cheapest path from s to u.*

2.4 Answers to selected problems

Problem 2.2. A leaf is a vertex with one outgoing edge; suppose there is no leaf. Pick a vertex, take one of its outgoing edges. As each vertex has at least two adjacent edges, we keep going arriving at one edge, and leaving by the other. As there are finitely many edges we must eventually form a cycle. Contradiction.

We now show by induction on n that a tree with n nodes must have exactly $n - 1$ edges. Basis case: $n = 1$, so the tree consists of a single node, and hence it has no edges; $n - 1 = 1 - 1 = 0$ edges. Induction step: suppose that we have a tree with $n + 1$ nodes. Pick a leaf and the edge that connects it to the rest of the tree. Removing this leaf and its edge results in a tree with n nodes, and hence—by induction hypothesis—with $n - 1$ edges. Thus, the entire tree has $(n - 1) + 1 = n$ edges, as required.

Problem 2.4. We prove this by induction, with the basis case $n = 3$ (since a graph—without multiple edges between the same pair of nodes—cannot have a cycle with less than 3 nodes). If $n = 3$, and there are more than $n - 1 = 2$ edges, there must be exactly 3 edges. So the graph is a cycle (a "triangle"). Induction step: consider a graph with $n + 1$ many nodes ($n \geq 3$), and at least $n + 1$ many edges. If the graph has a node with zero or one edges adjacent to it, then by removing that node (and its edge, if there is one), we obtain a graph with n nodes and at least n edges, and so—by induction hypothesis—the resulting graph has a cycle, and so the original graph also has a cycle. Otherwise, all nodes have at least two edges adjacent to it. Suppose v_0 is such a node, and $(v_0, x), (v_0, y)$ are two edges. Remove v_0 from the graph, and remove the edges $(v_0, x), (v_0, y)$ and replace them by the single edge (x, y). Again—by induction hypothesis—there must be a cycle in the resulting graph. But then there must be a cycle in the original graph as well. (Note that there are $n + 1$ nodes, so after removing v_0 there are n nodes, and $n \geq 3$.)

Problem 2.10. Let e be any edge in $T_2 - T_1$. We must prove the existence of $e' \in T_1 - T_2$ such that $(T_1 \cup \{e\}) - \{e'\}$ is a spanning tree. Since $e \notin T_1$, by adding e to T_1 we obtain a cycle (by lemma 2.3, which is proved in problem 2.4). A cycle has at least 3 edges (the graph G has at least 3 nodes, since otherwise it could not have two distinct spanning trees!). So in this cycle, there is an edge e' not in T_2. The reason is that if every edge e' in the cycle did belong to T_2, then T_2 itself would have a cycle. By removing e', we break the cycle but the resulting graph, $(T_1 \cup \{e\}) - \{e'\}$, is still connected and of size $|T_1| = |T_2|$, i.e., the right size for a tree, so it must be acyclic (for otherwise, we could get rid of some edge, and have a spanning tree of size smaller than T_1 and T_2—but all spanning trees have the same size), and therefore $(T_1 \cup \{e\}) - \{e'\}$ is a spanning tree.

Problem 2.12. Let T be any MCST for a graph G. Reorder the edges of G by costs, just as in Kruskal's algorithm. For any block of edges of the same cost, put those edges which appear in T before all the other edges in that block. Now prove the following loop invariant: the set of edges S

selected by the algorithm with the initial ordering as described is always a subset of T. Initially $S = \emptyset \subseteq T$. In the induction step, $S \subseteq T$, and S' is the result of adding one more edge to S. If $S' = S$ there is nothing to do, and if $S' = S \cup \{e\}$, then we need to show that $e \in T$. Suppose that it isn't. Let T' be the result of Kruskal's algorithm, which we know to be a MCST. By the exchange lemma, we know that there exists an $e' \notin T'$ such that $T \cup \{e\} - \{e'\}$ is a ST, and since T was a MCST, then $c(e') \leq c(e)$, and hence e' was considered *before* e. Since e' is not in T', it was rejected, so it must have created a cycle in S, and hence in T—contradiction. Thus $S \cup \{e\} \subseteq T$.

Problem 2.16. Since $S' = S$ and $S'_{opt} = S_{opt}$ we must show that S is extendable into S_{opt} with jobs in $\{i+1, i+2, \ldots, n\}$. Since job i could not be scheduled in S, and S_{opt} extends S (i.e., S_{opt} has all the jobs that S had, and perhaps more), it follows that i could not be in S_{opt} either, and so i was not necessary in extending S into S_{opt}.

Problem 2.18. why 1. To show that S'_{opt} is feasible, we have to show that no job is scheduled twice, and no job is scheduled after its deadline. The first is easy, because S_{opt} was feasible. For the second we argue like this: the job that was at time t_0 is now moved to $t_1 < t_0$, so certainly if t_0 was before its deadline, so is t_1. The job that was at time t_1 (job i) has now been moved forward to time t_0, but we are working under the assumption that job i was scheduled (at this point) in slot t_0, so $t_0 \leq d_i$, and we are done. **why 2.** S'_{opt} extends S' because S_{opt} extended S, and the only difference is positions t_1 and t_0. They coincide in position t_0 (both have i), so we only have to examine position t_1. But $S(t_1) = 0$ since $S_{opt}(t_1) = i$, and S does not schedule job i at all. Since the only difference between S and S' is in position t_0, it follows that $S'(t_1) = 0$, so it does not matter what $S'_{opt}(t_1)$ is, it will extend S'. **why 3.** They schedule the same set of jobs, so they must have the same profit. **why 4.** Suppose $t_1 > t_0$. Since S_{opt} extends S, it follows that $S(t_1) = 0$. Since $S_{opt}(t_1) = i$, it follows that $t_1 \leq d_i$. But then, the algorithm would have scheduled i in t_1, not in t_0.

The fact that $j \neq 0$ is used in the last sentence of the proof of claim 2.17, where we conclude a contradiction from $S(t_2) = j \neq S_{opt}(t_2)$. If j were 0 then it could very well be that $S(t_2) = j = 0$ but $S_{opt}(t_2) \neq 0$.

Problem 2.20. Define a *promising list* to be one that can be extended to an optimal list of coins. Now show that L is promising is a loop invariant. Basis case: Initially, L is empty, so any optimal solution extends L. Hence L is promising. Induction step: Assume that L is promising, and show that L continues being promising after one more execution of the loop:

Suppose L is promising, and $s < N$. Let L' be the list that extends L to the optimal solution, i.e. $L, L' = L_{\text{opt}}$. Let x be the largest item in C such that $s + x \leq N$. **Case (a)** $x \in L'$. Then $L' = x, L''$, so that L, x can be extended to the optimal solution L_{opt} by L''. **Case (b)** $x \notin L'$. We show that this case is not possible. To this end we prove the following claim:

Claim: If $x \notin L'$, then there is a sub-list L_0 of L' such that $x =$ sum of elements in L_0.

Proof of claim: Let B be the smallest number such that $B \geq x$, and some sub-list of L' sums to B. Let this sub-list be $\{e_1, e_2, \ldots, e_l\}$, where $e_i \leq e_{i+1}$ (i.e. the elements are in non-decreasing order). Since x is the largest coin that fits in $N - s$, and the sum of the coins in L' is $N - s$, it follows that every coin in L' is $\leq x$. Since $e_l \neq x$ (as $x \notin L'$), it follows that $l > 1$. Let $D = x - (e_2 + \ldots + e_l)$. By definition of B we know that $D > 0$. Each of the numbers $x, e_2, \ldots, e_l$ is divisible by e_1 (to see this note that all the coins are powers of p, i.e. in the set $\{1, p, p^2, \ldots, p^n\}$, and $e_l < x$ so $e_1 < x$). Thus $D \geq e_1$. On the other hand $x \leq e_1 + e_2 + \ldots + e_l$, so we also know that $D \leq e_1$, so in fact $D = e_1$. Therefore $x = e_1 + e_2 + \ldots + e_l$, and we are done. (end proof of claim)

Thus $\{e_1, e_2, \ldots, e_l\}$ can be replaced by the single coin x. If $l = 1$, then $x = e_1 \in L'$, which is a contradiction. If $l > 1$, then $L, x, L' - \{e_1, e_2, \ldots, e_l\}$ sums up to N, but it has less coins than $L, L' = L_{\text{opt}}$ which is a contradiction. Thus case (b) is not possible.

Problem 2.21. See algorithm 2.5

Algorithm 2.5 Solution to problem 2.21.

$w(e_1) \geq w(e_2) \geq \ldots \geq w(e_m)$
$M \longleftarrow \emptyset$
for $i : 1..m$ **do**
    if $M \cup \{e_i\}$ does *not* contain two edges with a common vertex
    then
        $M \longleftarrow M \cup \{e_i\}$
    end if
end for

Problem 2.23. Let M_{opt} be an optimal matching. Define "M is promising" to mean that M can be extended to M_{opt} with edges that have not been considered yet. We show that "M is promising" is a loop invariant of our algorithm. The result will follow from this (it will also follows that there is a *unique* max matching). Basis case: $M = \emptyset$, so it is certainly promising. Induction step: Assume M is promising, and let M' be M after considering edge e_i. We show that: $e_i \in M' \iff e_i \in M_{\text{opt}}$.

[$\implies$] Assume that $e_i \in M'$, since the weights are distinct, and powers of 2, $w(e_i) > \sum_{j=i+1}^{m} w(e_j)$ (to see why this holds, see problem 1.1), so unless $e_i \in M_{\text{opt}}$, $w(M_{\text{opt}}) < w$ where w is the result of algorithm.

[$\impliedby$] Assume that $e_i \in M_{\text{opt}}$, so $M \cup \{e_i\}$ has no conflict, so the algorithm would add it.

Problem 2.24. This problem refers Dijkstra's algorithm for the shortest path problem; for more background see §24.3, pg. 595, in [Cormen *et al.* (2001)] and §4.4, pg. 137, in [Kleinberg and Tardos (2006)].

2.5   Notes

Any book on algorithms has a chapter on greedy algorithms. For example, chapter 16 in [Cormen *et al.* (2001)] or chapter 4 in [Kleinberg and Tardos (2006)].

Another great example of a successful greedy solution are Huffman codes, which are a widely used and very effective technique for loss-less data compression. Huffman's algorithm uses a table of the frequencies of occurrences of the characters to build an optimal way of representing each character as a binary string. See §16.3 in [Cormen *et al.* (2001)].

Chapter 3

Divide and Conquer

Divide et impera—divide and conquer—was a Roman military strategy that consisted in securing command by breaking a large concentration of power into portions that alone were weaker, and methodically dispatching those portions one by one. This is the idea behind divide and conquer algorithms: take a large problem, divide it into smaller parts, solve those parts *recursively*, and combine the solutions to those parts into a solution to the whole.

The paradigmatic example of a divide and conquer algorithm is merge sort, where we have a large list of items to be sorted; we break it up into two smaller lists (divide), sort those recursively (conquer), and then combine those two sorted lists into one large sorted list. We present this algorithm in section 3.1. We also present a divide and conquer algorithm for binary integer multiplication—section 3.2, and graph reachability—section 3.3.

The divide and conquer approach is often used in situations where there is a brute force/exhaustive search algorithm that solves the problem, but the divide and conquer algorithm improves the running time. This is, for example, the case of binary integer multiplication. The last example in this chapter is a divide and conquer algorithm for reachability (Savitch's algorithm) that minimizes the use of memory, rather than the running time.

In order to analyze the use of resources (whether time or space) of a recursive procedure we must solve recurrences; see, for example, [Rosen (2007)] or [Cormen *et al.* (2001)] for the necessary background—the master method for solving recurrences. We provide a short discussion in the Notes section at the end of this chapter.

3.1 Mergesort

Suppose that we have two lists of numbers that are already sorted. That is, we have a list $a_1 \leq a_2 \leq \cdots \leq a_n$ and $b_1 \leq b_2 \leq \cdots \leq b_m$. We want to combine those two lists into one long sorted list $c_1 \leq c_2 \leq \cdots \leq c_{n+m}$. Algorithm 3.1 does the job.

Algorithm 3.1 Merge two lists

Pre-condition: $a_1 \leq a_2 \leq \cdots \leq a_n$ and $b_1 \leq b_2 \leq \cdots \leq b_m$

    $p_1 \longleftarrow 1; p_2 \longleftarrow 1; i \longleftarrow 1$
    while $i \leq n + m$ **do**
        if $a_{p_1} \leq b_{p_2}$ **then**
            $c_i \longleftarrow a_{p_1}; p_1 \longleftarrow p_1 + 1$
        else
            $c_i \longleftarrow b_{p_1}; p_2 \longleftarrow p_2 + 1$
        end if
        $i \longleftarrow i + 1$
    end while
Post-condition: $c_1 \leq c_2 \leq \cdots \leq c_{n+m}$

The mergesort algorithm sorts a given list of numbers by first dividing them into two equal halves (or, if the list is of odd-length, into two lists of length $\lceil n/2 \rceil$ and $\lfloor n/2 \rfloor$, respectively), sorting each half recursively, and then combining the results using algorithm 3.1.

Algorithm 3.2 Mergesort

Pre-condition: A list of numbers $a_1, a_2, \ldots, a_n$

1: $L \longleftarrow a_1, a_2, \ldots, a_n$
2: **if** $|L| \leq 1$ **then**
3:     **return** L
4: **else**
5:     $L_1 \longleftarrow$ first $\lceil n/2 \rceil$ elements of L
6:     $L_2 \longleftarrow$ last $\lfloor n/2 \rfloor$ elements of L
7:     **return** Merge(Mergesort(L_1), Mergesort(L_2))
8: **end if**

In algorithm 3.2, line 1 sets L to be the list of the input numbers $a_1, a_2, \ldots, a_n$. These are integers, not necessarily ordered. Line 2 checks if L is not empty or consists of a single element; if that is the case, then the list is already sorted—this is where the recursion bottoms out. Otherwise, in line 5 we let L_1 consist of the first $\lceil n/2 \rceil$ elements of L and L_2 consist of the last $\lfloor n/2 \rfloor$ elements of L.

Problem 3.1. *Show that* $L = L_1 \cup L_2$.

In section 1.3.5 of the Preliminaries chapter we have shown how to use the theory of fixed points to prove the correctness of recursive algorithms. For us this will remain a theoretical tool, as it is not easy to come up with the least fixed point that interprets a recursion. We are going to give natural proofs of correctness using induction.

Problem 3.2. *Prove the correctness of the Mergesort algorithm.*

Let $T(n)$ bound the running time of the mergesort algorithm on lists of length n. Clearly,

$$T(n) \leq T(\lceil n/2 \rceil) + T(\lfloor n/2 \rfloor) + cn,$$

where cn, for some constant c, is the cost of the merging of the two lists (algorithm 3.1). Furthermore, the asymptotic bounds are not affected by the floors and the ceils, and so we can simply say that $T(n) \leq 2T(n/2) + cn$. Thus, $T(n)$ is bounded by $O(n \log n)$.

3.2 Multiplying numbers in binary

Consider the example of multiplication of two binary numbers, using the junior school algorithm, given in figure 3.1.

	1	2	3	4	5	6	7	8
x					1	1	1	0
y					1	1	0	1
s_1					1	1	1	0
s_2				0	0	0	0	
s_3			1	1	1	0		
s_4		1	1	1	0			
$x \times y$	1	0	1	1	0	1	1	0

Fig. 3.1   Multiply 1110 times 1101.

This school multiplication algorithm is very simple. To multiply x times y, where x, y are two numbers in binary, we go through y from right to left;

when we encounter a 0 we write a row of as many zeros as $|x|$, the length of x. When we encounter a 1 we copy x. When we move to the next bit of y we shift by one space to the left. At the end we produce the familiar "stairs" shape—see s_1, s_2, s_3, s_4 in figure 3.1 (henceforth, figure 3.1 is our running example of binary multiplication).

Once we obtain the "stairs," we go back to the top step (line s_1) and to its right-most bit (column 8). To obtain the product we add all the entries in each column with the usual carry operation. For example, column 5 contains two ones, so we write a 0 in the last row (row $x \times y$) and carry over 1 to column 4. It is not hard to see that multiplying two n-bit integers takes $O(n^2)$ primitive bit operations.

We now present a divide and conquer algorithm that takes only $O(n^{\log 3}) \approx O(n^{1.59})$ operations. The speed-up obtained from the divide and conquer procedure appears slight—but the improvement does become substantial as n grows very large.

Let x and y be two n-bit integers. We break them up into two smaller $n/2$-bit integers as follows:

$$x = (x_1 \cdot 2^{n/2} + x_0),$$
$$y = (y_1 \cdot 2^{n/2} + y_0).$$

Thus x_1 and y_1 correspond to the high-order bits of x and y, respectively, and x_0 and y_0 to the low-order bits of x and y, respectively. The product of x and y appears as follows in terms of those parts:

$$xy = (x_1 \cdot 2^{n/2} + x_0)(y_1 \cdot 2^{n/2} + y_0)$$
$$= x_1 y_1 \cdot 2^n + (x_1 y_0 + x_0 y_1) \cdot 2^{n/2} + x_0 y_0. \tag{3.1}$$

A divide and conquer procedure appears surreptitiously. To compute the product of x and y we compute the four products $x_1 y_1, x_1 y_0, x_0 y_1, x_0 y_0$, *recursively*, and then we combine them as in (3.1) to obtain xy.

Let $T(n)$ be the number of operations that are required to compute the product of two n-bit integers using the divide and conquer procedure that arises from (3.1). Then

$$T(n) \le 4T(n/2) + cn, \tag{3.2}$$

since we have to compute the four products $x_1 y_1, x_1 y_0, x_0 y_1, x_0 y_0$ (this is where the $4T(n/2)$ factor comes from), and then we have to perform three additions of n-bit integers (that is where the factor cn, where c is some constant, comes from). Notice that we do not take into account the product by 2^n and $2^{n/2}$ (in (3.1)) as they simply consist in shifting the binary string by an appropriate number of bits to the left (n for 2^n and $n/2$ for $2^{n/2}$). These shift operations are inexpensive, and can be ignored in the complexity analysis.

When we solve the standard recurrence given by (3.2), we can see that $T(n) \leq O(n^{\log 4}) = O(n^2)$, so it seems that we have gained nothing over the brute force procedure.

It appears from (3.1) that we have to make four recursive calls; that is, we need to compute the four multiplications $x_1 y_1, x_1 y_0, x_0 y_1, x_0 y_0$. But we can get away with only three three multiplications, and hence four recursive calls: $x_1 y_1, x_0 y_0$ and $(x_1 + x_0)(y_1 + y_0)$; the reason being that

$$(x_1 y_0 + x_0 y_1) = (x_1 + x_0)(y_1 + y_0) - (x_1 y_1 + x_0 y_0). \tag{3.3}$$

See figure 3.2 for a comparison of the cost of operations.

	multiplications	additions	shifts
Method (3.1)	4	3	2
Method (3.3)	3	4	2

Fig. 3.2   The first line, labeled "Method (3.1)," shows the number of operations required when performing binary multiplication according to the recursion given by (3.1). The second line, labeled "Method (3.3)," shows the method that arises from using (3.3) to reduce the number of multiplications, which is also the number of recursive calls, by one with respect to the obvious procedure given by (3.1). It can be seen that reducing the number of multiplications by one increase the number of additions by one—something has to give. But, as multiplications are more expensive, the trade is worth it.

Algorithm 3.3 implements the idea given by (3.3).

Algorithm 3.3 Recursive binary multiplication

Pre-condition: Two n-bit integers x and y
1: $x \longleftarrow (x_1 \cdot 2^{\lceil n/2 \rceil} + x_0)$
2: $y \longleftarrow (y_1 \cdot 2^{\lceil n/2 \rceil} + y_0)$
3: $z_1 \longleftarrow \text{Multiply}(x_1 + x_0, y_1 + y_0)$
4: $z_2 \longleftarrow \text{Multiply}(x_1, y_1)$
5: $z_3 \longleftarrow \text{Multiply}(x_0, y_0)$
6: **return** $z_2 \cdot 2^n + (z_1 - z_2 - z_3) \cdot 2^{\lceil n/2 \rceil} + z_3$

Note that in lines 1 and 2 of the algorithm, we break up x and y into two parts x_1, x_0 and y_1, y_0, respectively, where x_1, y_1 consist of the $\lfloor n/2 \rfloor$ high order bits, and x_0, y_0 consist of the $\lceil n/2 \rceil$ low order bits.

Problem 3.3. *Prove the correctness of algorithm 3.3.*

Algorithm 3.3 takes $T(n) \leq 3T(n/2) + cn$ bit operations, where the factor of 3 comes from the 3 recursive calls (lines 3,4,5), the $n/2$ comes from dividing the string in two halves, and cn accounts for the additions and shifts performed on lines 3 and 6, for some constant c. Thus, the running time is $O(n^{\log 3}) \approx O(n^{1.59})$—to see this read the discussion on solving recurrences in the Notes section of this chapter.

3.3   Savitch's algorithm

In this section we are going to give a divide and conquer solution to the graph reachability problem. Recall the graph-theoretic definitions that were given at the beginning of section 2.1. Here we assume that we have a (directed) graph G, and we want to establish whether there is a path from some node s to some node t; note that we are not even searching for a shortest path (as in section 2.3.3 or in section 4.2); we just want to know if node t is reachable from node s.

As a twist on minimizing the running time of algorithms, we are going to present a very clever divide and conquer solution, due to Walter Savitch, that minimizes the amount of space (i.e., memory) used by the algorithm. Savitch's algorithm solves (directed) reachability in space $O(\log^2 n)$, where n is the number of vertices of the graph. This is remarkable, as $O(\log^2 n)$ bits of memory is very little space indeed, for a graph with n vertices! We assume that the graph is presented as an $n \times n$ adjacency matrix, and so it takes exactly n^2 bits of memory.

Define the Boolean predicate $R(G, u, v, i)$ to be true iff there is a path in G from u to v of length at most 2^i. The key observation is that if a path exists from u to v, then any such path must have a mid-point w; a seemingly trivial observation that nevertheless inspires a very clever recursive procedure. In other words there exist paths of distance at most 2^{i-1} from u to w and from w to v, i.e.,

$$R(G, u, v, i) \iff (\exists w)[R(G, u, w, i-1) \wedge R(G, w, v, i-1)]. \qquad (3.4)$$

Algorithm 3.4 computes the predicate $R(G, u, v, i)$ based on the recurrence given in (3.4). Note that in algorithm 3.4 we are computing $R(G, u, v, i)$; the recursive calls come in line 9 where we compute $R(G, u, w, i-1)$ and $R(G, w, v, i-1)$.

Algorithm 3.4 Savitch

1: **if** $i = 0$ **then**
2:     **if** $u = v$ **then**
3:         **return** T
4:     **else if** (u, v) is an edge **then**
5:         **return** T
6:     **end if**
7: **else**
8:     **for** every vertex w **do**
9:         **if** R$(G, u, w, i - 1)$ and R$(G, w, v, i - 1)$ **then**
10:             **return** T
11:         **end if**
12:     **end for**
13: **end if**
14: **return** F

Problem 3.4. *Show that algorithm 3.4 is correct (i.e., it computes* R(G, u, v, i) *correctly) and it requires at most* $i \cdot s$ *space, where s is the number of bits required to keep record of a single node. Conclude that it requires* $O(\log^2 n)$ *space on a graph G with n nodes.*

Problem 3.5. *Algorithm 3.4 truly uses very little space to establish connectivity in a graph. But what is the* time *complexity of this algorithm?*

3.4 Answers to selected problems

Problem 3.1. It is enough to show that $\lceil n/2 \rceil + \lfloor n/2 \rfloor = n$. If n is even, then $\lceil n/2 \rceil = \lfloor n/2 \rfloor = n/2$, and $n/2 + n/2 = n$. If n is odd, then $n = 2k+1$, and so $\lceil n/2 \rceil = k + 1$ while $\lfloor n/2 \rfloor = k$ and $(k + 1) + k = 2k + 1 = n$.

Problem 3.2. We must show that given a list of integers $L = a_1, a_2, \ldots, a_n$, the algorithm returns a list L', which consists of the numbers in L in nondecreasing order. The recursion itself suggest the right induction; we use the CIP (see page 2). If $|L| = 1$ then $L' = L$ and we are done. Otherwise, $|L| = n > 1$, and we obtain two lists L_1 and L_2 (of lengths $\lceil n/2 \rceil$ and $n - \lceil n/2 \rceil$), which, by induction hypothesis, are returned ordered. Now it remains to prove the correctness of the merging procedure, algorithm 3.1, which can also be done by induction.

Problem 3.5. $O(2^{\log^2 n}) = O(n^{\log n})$, so the time complexity of Savitch's algorithm is super-polynomial, and so not very good.

3.5   Notes

For a full discussion of Mergesort and binary multiplication, see §5.1 and §5.5, respectively, in [Kleinberg and Tardos (2006)]. For more background on Savitch's algorithm (section 3.3) see theorem 7.5 in [Papadimitriou (1994)], §8.1 in [Sipser (2006)] or theorem 2.7 in [Kozen (2006)].

See chapter 7 in [Rosen (2007)] for an introduction to solving recurrence relations, and §4.3 in [Cormen et al. (2001)] for a very thorough discussion of the *Master method for solving recurrences*. Here we include a very short discussion; we want to solve recurrences of the following form:

$$T(n) = aT(n/b) + f(n), \tag{3.5}$$

where $a \geq 1$ and $b > 1$ are constants and $f(n)$ is an asymptotically positive function—meaning that there exists an n_0 such that $f(n) > 0$ for all $n \geq n_0$. There are three cases for solving such a recurrence.

Case 1 is $f(n) = O(n^{\log_b a - \varepsilon})$ for some constant $\varepsilon > 0$; in this case we have that $T(n) = \Theta(n^{\log_b a})$. Case 2 is $f(n) = \Theta(n^{\log_b a} \log^k n)$ with $k \geq 0$; in this case we have that $T(n) = \Theta(n^{\log_b a} \log^{k+1} n)$. Finally, Case 3 is $f(n) = \Omega(n^{\log_b a + \varepsilon})$ with $\varepsilon > 0$, and $f(n)$ satisfies the *regularity condition*, namely $af(n/b) \leq cf(n)$ for some constant $c < 1$ and all sufficiently large n; in this case $T(n) = \Theta(f(n))$.

For example, the recurrence that appears in the analysis of mergesort is $T(n) = 2T(n/2) + cn$, so $a = 2$ and $b = 2$, and so $\log_b a = \log_2 2 = 1$, and so we can say that $f(n) = \Theta(n^{\log_b a} \log^k n) = \Theta(n \log n)$, i.e., $k = 1$ in Case 2, and so $T(n) = \Theta(n \log n)$ as was pointed out in the analysis.

Chapter 4

Dynamic Programming

Dynamic programming is an algorithmic technique that is closely related to the divide and conquer approach we saw in the previous chapter. However, while the divide and conquer approach is essentially recursive, and so "top down," dynamic programming works "bottom up."

A dynamic programming algorithm creates an array of related but simpler subproblems, and then, it computes the solution to the big complicated problem by using the solutions to the easier subproblems which are stored in the array. We usually want to maximize profit or minimize cost.

There are three steps in finding a dynamic programming solution to a problem: (i) Define a class of subproblems, (ii) give a recurrence based on solving each subproblem in terms of simpler subproblems, and (iii) give an algorithm for computing the recurrence.

4.1 Longest monotone subsequence problem

Input: $d, a_1, a_2, \ldots, a_d \in \mathbb{N}$.
Output: $L =$ length of the longest monotone non-decreasing subsequence.

Note that a subsequence need not be consecutive, that is $a_{i_1}, a_{i_2}, \ldots, a_{i_k}$ is a monotone subsequence provided that

$$1 \leq i_1 < i_2 < \ldots < i_k \leq d,$$
$$a_{i_1} \leq a_{i_2} \leq \ldots \leq a_{i_k}.$$

For example, the length of the longest monotone subsequence (henceforth LMS) of $\{4, 6, 5, 9, 1\}$ is 3.

We first define an array of subproblems: $R(j) =$ length of the longest monotone subsequence which ends in a_j. The answer can be extracted from array R by computing $L = \max_{1 \leq j \leq n} R(j)$.

The next step is to find a recurrence. Let $R(1) = 1$, and for $j > 1$,

$$R(j) = \begin{cases} 1 & \text{if } a_i > a_j \text{ for all } 1 \leq i < j \\ 1 + \max_{1 \leq i < j}\{R(i) | a_i \leq a_j\} & \text{otherwise} \end{cases}.$$

We finish by writing an algorithm that computes R; see algorithm 4.1.

Algorithm 4.1 Longest monotone subsequence (LMS)

$R(1) \leftarrow 1$
for $j : 2..d$ **do**
     max $\leftarrow 0$
     for $i : 1..j - 1$ **do**
         if $R(i) >$ max and $a_i \leq a_j$ **then**
             max $\leftarrow R(i)$
         end if
     end for
     $R(j) \leftarrow$ max $+1$
end for

Problem 4.1. *Once we have computed all the values of the array R, how could we build an actual monotone non-decreasing subsequence of length L?*

Problem 4.2. *What would be the appropriate pre/post-conditions of the above algorithms? Prove correctness with an appropriate loop invariant.*

Problem 4.3. *Consider the following variant of the Longest Monotone Subsequence problem. The input is still $d, a_1, a_2, \ldots, a_d \in \mathbb{N}$, but the output is the the length of the longest subsequence of $a_1, a_2, \ldots, a_d$, where any two consecutive members of the subsequence differ by at most 1. For example, the longest such subsequence of $\{7, 6, 1, 4, 7, 8, 20\}$ is $\{7, 6, 7, 8\}$, so in this case the answer would be 4. Give a dynamic programming solution to this problem.*

4.2   All pairs shortest path problem

Input: Directed graph $G = (V, E)$, $V = \{1, 2, \ldots, n\}$, and a cost function $C(i, j) \in \mathbb{N}^+ \cup \{\infty\}$, $1 \leq i, j \leq n$, $C(i, j) = \infty$ if (i, j) is not an edge.
Output: An array D, where $D(i, j)$ the length of the shortest directed path from i to j.

Recall that we have defined undirected graphs in section 2.1; a *directed graph* (or *digraph*) is a graph where the edges have a direction, i.e., the edges are arrows. Also recall that in section 2.3.3 we have given a greedy algorithm for computing the shortest paths from a designated node s to all the nodes in an (undirected) graph.

Problem 4.4. *Construct a family of graphs $\{G_n\}$, where G_n has $O(n)$ many nodes, and exponentially many paths, that is $\Omega(2^n)$ paths. Conclude, therefore, that an exhaustive search is not a feasible solutions to the "all pairs shortest path problem."*

Define an array of subproblems: let $A(k, i, j)$ be the length of the shortest path from i to j such that all *intermediate* nodes on the path are in $\{1, 2, \ldots, k\}$. Then $A(n, i, j) = D(i, j)$ will be the solution. The convention is that if $k = 0$ then $\{1, 2, \ldots, k\} = \emptyset$.

Define a recurrence: we first initialize the array for $k = 0$ as follows: $A(0, i, j) = C(i, j)$. Now we want to compute $A(k, i, j)$ for $k > 0$. To design the recurrence, notice that the shortest path between i and j either includes k or does not. Assume we know $A(k - 1, r, s)$ for all r, s. Suppose node k is not included. Then, obviously, $A(k, i, j) = A(k - 1, i, j)$. If, on the other hand, node k occurs on a shortest path, then it occurs exactly once, so $A(k, i, j) = A(k - 1, i, k) + A(k - 1, k, j)$. Therefore, the shortest path length is obtained by taking the minimum of these two cases:

$$A(k, i, j) = \min\{A(k - 1, i, j), A(k - 1, i, k) + A(k - 1, k, j)\}.$$

Write an algorithm: it turns out that we only need space for a two-dimensional array $B(i, j) = A(k, i, j)$, because to compute $A(k, *, *)$ from $A(k - 1, *, *)$ we can overwrite $A(k - 1, *, *)$.

Our solution is algorithm 4.2, known as Floyd's algorithm (or the Floyd-Warshall algorithm). It is remarkable as it runs in time $O(n^3)$, where n is the number of vertices, while there may be up to $O(n^2)$ edges in such a graph. In lines 1–5 we initialize the array B, i.e., we set it equal to C. Note that before line 6 is executed, it is the case that $B(i, j) = A(k - 1, i, j)$ for all i, j.

Problem 4.5. *Why does the overwriting method in algorithm 4.2 work? The worry is that $B(i, k)$ or $B(k, j)$ may have already been updated (if $k < j$ or $k < i$). However, the overwriting does work; explain why. We could have avoided a 3-dimensional array by keeping two 2-dimensional arrays instead, and then overwriting would not be an issue at all; how would that work?*

Algorithm 4.2 Floyd

1: **for** $i : 1..n$ **do**
2:        **for** $j : 1..n$ **do**
3:                $B(i,j) \longleftarrow C(i,j)$
4:        **end for**
5: **end for**
6: **for** $k : 1..n$ **do**
7:        **for** $i : 1..n$ **do**
8:                **for** $j : 1..n$ **do**
9:                        $B(i,j) \longleftarrow \min\{B(i,j), B(i,k) + B(k,j)\}$
10:                **end for**
11:        **end for**
12: **end for**
13: **return** $D \longleftarrow B$

Problem 4.6. *What are appropriate pre and post-conditions? What is an appropriate loop invariant?*

4.3   Simple knapsack problem

Input: $w_1, w_2, \ldots, w_d, C \in \mathbb{N}$, where C is the knapsack's capacity.
Output: $\max_S \{K(S) | K(S) \leq C\}$, where $S \subseteq [d]$ and $K(S) = \sum_{i \in S} w_i$.

This is an NP-hard[1] problem, which means that we cannot expect to find a polynomial time algorithm that works in general. We give a dynamic programming solution that works for relatively small C; note that for our method to work the inputs $w_1, \ldots, w_d, C$ must be (non-negative) integers. We often abbreviate the name "simple knapsack problem" with SKS.

Define an array of subproblems: we consider the first i weights (i.e., $[i]$) summing up to an *intermediate* weight limit j. We define a Boolean array R as follows:

$$R(i,j) = \begin{cases} \mathsf{T} & \text{if } \exists S \subseteq [i] \text{ such that } K(S) = j \\ \mathsf{F} & \text{otherwise} \end{cases},$$

---

[1]NP is the class of problems solvable in polynomial time on a *non-deterministic* Turing machine. A problem P is NP-hard if every problem in NP is reducible to P in polynomial time, that is, every problem in NP can be efficiently restate in terms of P. When a problem is NP-hard this is an indication that it is probably *intractable*, i.e., it cannot be solved efficiently in general. For more information on this see any book on complexity, for example [Papadimitriou (1994); Sipser (2006); Soltys (2009)].

for $0 \leq i \leq d$ and $0 \leq j \leq C$. Once we have computed all the values of R we can obtain the solution M as follows: $M = \max_{j \leq C}\{j|R(d,j) = \mathsf{T}\}$.

Define a recurrence: we initialize $R(0,j) = \mathsf{F}$ for $j = 1, 2, \ldots, C$, and $R(i,0) = \mathsf{T}$ for $i = 0, 1, \ldots, d$.

We now define the recurrence for computing R, for $i, j > 0$, in a way that hinges on whether we include object i in the knapsack. Suppose that we do *not* include object i. Then, obviously, $R(i,j) = \mathsf{T}$ iff $R(i-1,j) = \mathsf{T}$. Suppose, on the other hand, that object i *is* included. Then it must be the case that $R(i,j) = \mathsf{T}$ iff $R(i-1, j - w_i) = \mathsf{T}$ and $j - w_i \geq 0$, i.e., there is a subset $S \subseteq [i-1]$ such that $K(S)$ is exactly $j - w_i$ (in which case $j \geq w_i$). Putting it all together we obtain the following recurrence for $i, j > 0$:

$$R(i,j) = \mathsf{T} \iff R(i-1, j) = \mathsf{T} \vee (j \geq w_i \wedge R(i-1, j - w_i) = \mathsf{T}). \quad (4.1)$$

R	0	$\cdots$	$j{-}w_i$	$\cdots$	j	$\cdots$	C
0	T	F$\cdots$F	F	F$\cdots$F	F	F$\cdots$F	F
	T						
$\vdots$	$\vdots$						
	T						
$i{-}1$	T		**c**		**b**		
i	T				**a**		
	T						
$\vdots$	$\vdots$						
	T						
d	T						

Fig. 4.1 The recurrence given by the equivalence (4.1) can be interpreted as follows: we place a T in the square labeled with **a** if and only if at least one of the following two conditions is satisfied: there is a T in the position right above it, i.e., in the square labeled with **b** (if we can construct j with the first $i-1$ weights, surely we can construct j with he first i weights), or there is a T in the square labeled with **c** (if we can construct $j - w_i$ with the first $i - 1$ weights, surely we can construct j with the first i weights). Also note that to fill the square labeled with a we only need to look at two squares, and neither of those two squares is to the right; this will be important in the design of the algorithm (algorithm 4.3).

We finally design algorithm 4.3 that uses the same space saving trick as algorithm 4.2; it employs a one-dimensional array $S(j)$ for keeping track of a two-dimensional array $R(i,j)$. This is done by overwriting $R(i,j)$ with $R(i+1, j)$.

Algorithm 4.3 Simple knapsack (SKS)

1: $S(0) \longleftarrow \mathsf{T}$
2: **for** $j : 1..C$ **do**
3:        $S(j) \longleftarrow \mathsf{F}$
4: **end for**
5: **for** $i : 1..d$ **do**
6:        **for** *decreasing* $j : C..1$ **do**
7:              **if** $(j \geq w_i$ and $S(j - w_i) = \mathsf{T})$ **then**
8:                  $S(j) \longleftarrow \mathsf{T}$
9:              **end if**
10:        **end for**
11: **end for**

In algorithm 4.3, in line 1 we initialize the array for $i = j = 0$. In lines 2–4 we initialize the array for $i = 0$ and $j \in \{1, 2, \ldots, C\}$. Note that after each execution of the i-loop (line 5) it is the case that $S(j) = R(i, j)$ for all j.

Problem 4.7. *We are using a one dimensional array to keep track of a two dimensional array, but the overwriting is not a problem; explain why.*

Problem 4.8. *The assertion $S(j) = R(i, j)$ can be proved by induction on the number of times the i-loop in algorithm 4.3 is executed. This assertion implies that upon termination of the algorithm, $S(j) = R(d, j)$ for all j. Prove this formally, by giving pre/post-conditions, a loop invariant, and a standard proof of correctness.*

Problem 4.9. *Construct an input for which algorithm 4.3 would make an error if the inner loop "for decreasing $j : C..1$" (line 6) were changed to "for $j : 1..C$."*

Algorithm 4.3 is a nice illustration of the powerful idea of *program refinement*. We start with the idea of computing $R(i, j)$ for all i, j. We then realize that we only really need two rows in memory; to compute row i we only need to look up row $i - 1$. We then take it further and see that by updating row i from right to left we do not require row $i - 1$ at all—we can do it *mise en place*. By starting with a robust idea, and by successively trimming the unnecessary fat, we obtain a slick solution.

But how slick is our dynamic programming solution in terms of the complexity of the problem? That is, how many steps does it take to compute the solution *proportionally* to the size of the input? We must construct a $d \times C$ table and fill it in, so the time complexity of our solution is $O(d \cdot C)$. This seems acceptable at first glance, but we were saying in the introduction to this section that SKS is an NP-hard problem; what gives?

The point is that the input is assumed to be given in binary, and to encode C in binary we require only $\log C$ bits, and so the number of columns (C) is in fact exponential in the size of the input ($C = 2^{\log C}$). On the other hand, d is the number of weights, and since those weights must be listed somehow, the size of the list of weights is certainly bigger than d (i.e., this list cannot be encoded—in general—with $\log d$ bits; it requires at least d bits).

All we can say is that if C is of size $O(d^k)$, for some constant k, then our dynamic programming solution works in polynomial time in the size of the input. In other words, we have an efficient solution for "small" values of C. Another way of saying this is that as long as $|C|$ (the size of the binary encoding of C) is $O(\log d)$ our solution works in polynomial time.

Problem 4.10. *Show how to construct the actual optimal set of weights once R has been computed.*

Problem 4.11. *Define a "natural" greedy algorithm for solving SKS; let $\overline{M}$ be the output of this algorithm, and let M be the output of the dynamic programming solution given in this section. Show that either $\overline{M} = M$ or $\overline{M} > \frac{1}{2}C$.*

This last problem, problem 4.11, surreptitiously introduces the important concept of *approximation algorithms*. As was mentioned at the beginning of this section (see footnote on page 54), SKS is an example of an NP-hard problem, a problem to which we suspect there may be no efficient solution in the general case. That is, the majority of experts believe that any algorithm—attempting to solve SKS in the general case—will necessarily take an inordinate number of steps (i.e., time) to produce a solution (on some inputs at least).

One possible compromise is to design an efficient algorithm that does not give an *optimal* solution—which may not even be required—but only a solution with some guarantees as to its closeness to an optimal solution. Thus, we merely *approximate* the optimal solution but at least our algorithm runs quickly. The study of such compromises is undertaken by the field of approximation algorithms.

Finally, in the section below we give a greedy solution to SKS in the particular case where the weights have a certain "increasing property." This is an example of a *promise* problem, where we can expect some convenient condition on the inputs; a condition that we need not check for, but assume that we have. Note that we have been using the term "promising" to prove the correctness of greedy algorithms—this is a different notion from that of a "promise problem."

4.3.1 *Dispersed knapsack*

Input: $w_1, \ldots, w_d, C \in \mathbb{N}$, such that $w_i \geq \sum_{j=i+1}^{d} w_j$ for $i = 1, \ldots, d - 1$.
Output: $S_{\max} \subseteq [d]$ where $K(S_{\max}) = \max\limits_{S \subseteq [d]} \{K(S) | K(S) \leq C\}$.

Problem 4.12. *Give a "natural" greedy algorithm which solves Dispersed Knapsack by filling in the blanks in algorithm 4.4.*

Algorithm 4.4 Dispersed knapsack

$S \longleftarrow \emptyset$
for $i : 1..d$ **do**
      if _____ **then**

             _____

      end if
end for

Problem 4.13. *Give a definition of what it means for an intermediate solution S in algorithm 4.4 to be "promising." Show that the loop invariant "S is promising" implies that the greedy algorithm gives the optimal solution. Finally, show that "S is promising" is a loop invariant.*

4.4 General knapsack problem

Input: $w_1, w_2, \ldots, w_d, v_1, \ldots, v_d, C \in \mathbb{N}$
Output: $\max_{S \subseteq [d]} \{V(S) | K(S) \leq C\}$, $K(S) = \sum_{i \in S} w_i$, $V(S) = \sum_{i \in S} v_i$.

Thus, the general knapsack problem (which we abbreviate as GKS) has a positive integer value v_i besides each weight w_i, and the goal is to have as valuable a knapsack as possible, without exceeding C, i.e., the weight capacity of the knapsack.

More precisely, $V(S) = \sum_{i \in S} v_i$ is the total value of the set S of weights. The goal is to maximize $V(S)$, subject to the constraint that $K(S)$ (the sum of the weights in S) is at most C. Note that SKS is a special case of GKS where $v_i = w_i$, for all $1 \leq i \leq d$.

To solve GKS, we start by computing the same Boolean array $R(i,j)$ that was used to solve SKS. Thus $R(i,j)$ ignores the values v_i, and only depends on the weights w_i. Next we define another array $V(i,j)$ that depends on the values v_i as follows:

$$V(i,j) = \max\{V(S)|S \subseteq [i] \text{ and } K(S) = j\}, \qquad (4.2)$$

for $0 \leq i \leq d$ and $0 \leq j \leq C$.

Problem 4.14. *Give a recurrence for computing the array $V(i,j)$, using the Boolean array $R(i,j)$—assume that the array $R(i,j)$ has already been computed. Also, give an algorithm for computing $V(i,j)$.*

Problem 4.15. *If the definition of $V(i,j)$ given in (4.2) is changed so that we only require $K(S) \leq j$ instead of $K(S) = j$, then the Boolean array $R(i,j)$ is not needed in the recurrence. Give the recurrence in this case.*

4.5 Activity selection problem

Input: A list of activities $(s_1, f_1, p_1), \ldots, (s_n, f_n, p_n)$, where $p_i > 0$, $s_i < f_i$ and s_i, f_i, p_i are non-negative real numbers.
Output: A set $S \subseteq [n]$ of selected activities such that no two selected activities overlap, and the profit $P(S) = \sum_{i \in S} p_i$ is as large as possible.

An *activity i* has a fixed start time s_i, finish time f_i and profit p_i. Given a set of activities, we want to select a subset of non-overlapping activities with maximum total profit. A typical example of the activity selection problem is a set of lectures with fixed start and finish times that need to be scheduled in a single class room.

Define an array of subproblems: sort the activities by their finish times, $f_1 \leq f_2 \leq \ldots \leq f_n$. As it is possible that activities finish at the same time, we select the *distinct* finish times, and denote them $u_1 < u_2 < \ldots < u_k$, where, clearly, $k \leq n$. For instance, if we have activities finishing at times 1.24, 4, 3.77, 1.24, 5 and 3.77, then we partition them into four groups: activities finishing at times $u_1 = 1.24$, $u_2 = 3.77$, $u_3 = 4$, $u_4 = 5$.

Let u_0 be $\min_{1 \leq i \leq n} s_i$, i.e., the earliest start time. Thus,

$$u_0 < u_1 < u_2 < \ldots < u_k,$$

as it is understood that $s_i < f_i$. Define an array $A(0..k)$ as follows:

$$A(j) = \max_{S \subseteq [n]} \{P(S) | S \text{ is feasible and } f_i \le u_j \text{ for each } i \in S\},$$

where S is *feasible* if no two activities in S overlap. Note that $A(k)$ is the maximum possible profit for all feasible schedules S.

Problem 4.16. *Give a formal definition of what it means for a schedule of activities to be feasible, i.e., express precisely that the activities in a set S "do not overlap."*

Define a recurrence for $A(0..k)$. In order to give such a recurrence we first define an auxiliary array $H(1..n)$ such that $H(i)$ is the index of the largest distinct finish time no greater than the start time of activity i. Formally, $H(i) = l$ if l is the largest number such that $u_l \le s_i$. To compute $H(i)$, we need to search the list of distinct finish times. To do it efficiently, for each i, apply the binary search procedure that runs in logarithmic time in the length of the list of distinct finish times (try $l = \lfloor \frac{k}{2} \rfloor$ first). Since the length k of the list of distinct finish times is at most n, and we need to apply binary search for each element of the array $H(1..n)$, the time required to compute all entries of the array is $O(n \log n)$.

We initialize $A(0) = 0$, and we want to compute $A(j)$ given that we already have $A(0), \ldots, A(j-1)$. Consider $u_0 < u_1 < u_2 < \ldots < u_{j-1} < u_j$. Can we beat profit $A(j-1)$ by scheduling some activity that finishes at time u_j? Try all activities that finish at this time and compute maximum profit in each case. We obtain the following recurrence:

$$A(j) = \max\{A(j-1), \max_{1 \le i \le n}\{p_i + A(H(i)) \mid f_i = u_j\}\}, \qquad (4.3)$$

where $H(i)$ is the greatest l such that $u_l \le s_i$. Consider the example given in figure 4.2.

Consider the example in figure 4.3. To see how the bottom row of the right-hand table in figure 4.3 was computed, note that according to the recurrence (4.3), we have:

$$A(2) = \max\{20, 30 + A(0), 20 + A(1)\} = 40,$$
$$A(3) = \max\{40, 30 + A(0)\} = 40.$$

Therefore, the maximum profit is $A(3) = 40$.

Problem 4.17. *Write the algorithm.*

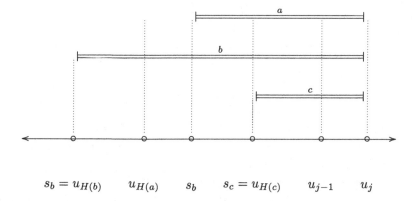

$$s_b = u_{H(b)} \quad u_{H(a)} \quad s_b \quad s_c = u_{H(c)} \quad u_{j-1} \quad u_j$$

Fig. 4.2 In this example we want to compute $A(j)$. Suppose that some activity finishing at time u_j must be scheduled in order to obtain the maximum possible profit. In this figure there are three activities that end at time u_j: a, b, c, given by $(s_a, f_a, p_a), (s_b, f_b, p_b), (s_c, f_c, p_c)$, respectively, where of course the assumption is that $u_j = f_a = f_b = f_c$. The question is which of these three activities must be selected. In order to establish this, we must look at each activity a, b, c in turn, and see what is the most profitable schedule that we can get if we *insist* that the given activity is scheduled. For example, if we insist that activity a be scheduled, we must see what is the most profitable schedule we can get where all other activities must finish by s_a, which effectively means that all other activities must finish by $u_{H(a)}$. Note that in this example we have that $u_{H(a)} < s_a$, but $u_{H(b)} = s_b$ and $u_{H(c)} = s_c$. When all is said, we must find which of the three values $p_a + A(H(a)), p_b + A(H(b)), p_c + A(H(c))$ is maximal.

Activity i:	1	2	3	4
Start s_i:	0	2	3	2
Finish f_i:	3	6	6	10
Profit p_i:	20	30	20	30
$H(i)$:	0	0	1	0

j:	0	1	2	3
u_j:	0	3	6	10
$A(j)$:	0	20	40	40

Fig. 4.3 An example with four activities.

Problem 4.18. *Given that A has been computed, how do you find a set of activities S such that $P(S) = A(k)$? Hint: If $A(k) = A(k-1)$, then we know that no selected activity finishes at time u_k, so we go on to consider $A(k-1)$. If $A(k) > A(k-1)$, then some selected activity finishes at time u_k. How do we find this activity?*

4.6   Jobs with deadlines, durations and profits

Input: A list of jobs $(d_1, t_1, p_1), \ldots, (d_n, t_n, p_n)$.
Output: A feasible schedule $C(1..n)$ such that the profit of C, denoted $P(C)$, is the maximum possible among feasible schedules.

In section 2.2 we considered the job scheduling problems for the case where each job takes unit time, i.e., each duration $d_i = 1$. We now generalize this to the case in which each job i has an arbitrary duration d_i, deadline t_i and profit p_i. We assume that d_i and t_i are positive integers, but the profit p_i can be a positive real number. We say that the schedule $C(1..n)$ is *feasible* if the following two conditions hold (let $C(i) = -1$ denote that job i is *not* scheduled, and so $C(i) \geq 0$ indicates that it it *is* scheduled, and note that we do allow jobs to be scheduled at time 0):

(1) if $C(i) \geq 0$, then $C(i) + d_i \leq t_i$ and
(2) if $i \neq j$ and $C(i), C(j) \geq 0$ then $C(i) + d_i \leq C(j)$ or $C(j) + d_j \leq C(i)$.

The first condition is akin to saying that each scheduled job finishes by its deadline and the second condition is akin to saying that no two scheduled jobs overlap. The goal is to find a feasible schedule $C(1..n)$ for the n jobs for which the profit $P(C) = \sum_{C(i) \geq 0} p_i$, the sum of the profits of the scheduled jobs, is maximized.

A job differs from an activity in that a job can be scheduled any time as long as it finishes by its deadline; an activity has a fixed start time and finish time. Because of the flexibility in scheduling jobs, it is "harder" to find an optimal schedule for jobs than to select an optimal subset of activities.

Note that job scheduling is "at least as hard as SKS." In fact an SKS instance $w_1, \ldots, w_n, C$ can be viewed as a job scheduling problem in which each duration $d_i = w_i$, each deadline $t_i = C$, and each profit $p_i = w_i$. Then the maximum profit of any schedule is the same as the maximum weight that can be put into the knapsack. This seemingly innocent idea of "at least as hard as" is in fact a powerful tool widely used in the field of computational complexity to compare the relative difficulty of problems. By restating a general instance of job scheduling as an instance of SKS we provided a *reduction* of job scheduling to SKS, and shown thereby that if one were able to solve job scheduling efficiently, one would automatically have an efficient solution to SKS.

To give a dynamic programming solution to the job scheduling problem, we start by sorting the jobs according to deadlines. Thus, we assume that

$t_1 \leq t_2 \leq \ldots \leq t_n$.

It turns out that to define a suitable array A for solving the problem, we must consider all possible integer times t, $0 \leq t \leq t_n$ as a deadline for the first i jobs. It is not enough to only consider the specified deadline t_i given in the problem input. Thus define the array $A(i, t)$ as follows:

$$A(i, t) = \max \left\{ P(C) : \begin{array}{l} C \text{ is a feasible schedule} \\ \text{only jobs in } [i] \text{ are scheduled} \\ \text{all scheduled jobs finish by time } t \end{array} \right\}.$$

We now want to design a recurrence for computing $A(i, t)$. In the usual style, consider the two cases that either job i occurs or does not occur in the optimal schedule (and note that job i will not occur in the optimal schedule if $d_i > \min\{t_i, t\}$). If job i does not occur, we already know the optimal profit.

If, on the other hand, job i does occur in an optimal schedule, then we may as well assume that it is the last job (among jobs $\{1, \ldots, i\}$) to be scheduled, because it has the latest deadline. Hence we assume that job i is scheduled as late as possible, so that it finishes either at time t, or at time t_i, whichever is smaller, i.e., it finishes at time $t_{\min} = \min\{t_i, t\}$.

Problem 4.19. *In light of the discussion in the above two paragraphs, find a recurrence for $A(i, t)$.*

4.7 Further examples and problems

4.7.1 *Consecutive subsequence sum problem*

Input: Real numbers $r_1, \ldots, r_n$

Output: For each consecutive subsequence of the form $r_i, r_{i+1}, \ldots, r_j$ let

$$S_{ij} = r_i + r_{i+1} + \cdots + r_j$$

where $S_{ii} = r_i$. Find $M = \max_{1 \leq i \leq j \leq n} S_{ij}$.

For example, in figure 4.4 we have a sample consecutive subsequence sum problem. There, the solution is $M = S_{35} = 3 + (-1) + 2 = 4$.

This problem can be solved in time $O(n^2)$ by systematically computing all of the sums S_{ij} and finding the maximum (there are $\binom{n}{2}$ pairs $i, j \leq n$ such that $i < j$). However, there is a more efficient dynamic programming solution which runs in time $O(n)$.

Define the array $M(1..n)$ by:

$$M(j) = \max\{S_{1j}, S_{2j}, \ldots, S_{jj}\}.$$

$$
\begin{array}{c|cccc}
j & 1 & 2\,3 & 4\,5 & 6\,7 \\
\hline
r_j & 1 & -5\,3 & -1\,2 & -8\,3 \\
M(j) & 1 & -4\,3 & 2\,4 & -4\,3
\end{array}
$$

Fig. 4.4   An example of computing $M(j)$.

See figure 4.4 for an example.

Problem 4.20. *Explain how to find the solution M from the array $M(1..n)$.*

Problem 4.21. *Complete the four lines indicated in algorithm 4.5 for computing the values of the array $M(1..n)$, given $r_1, r_2, \ldots, r_n$.*

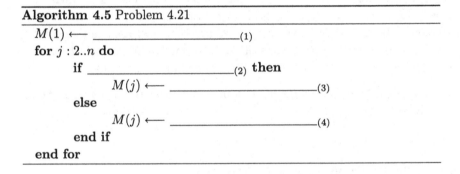

Algorithm 4.5 Problem 4.21

$M(1) \longleftarrow$ _____(1)

for $j : 2..n$ **do**

    if _____(2) **then**

        $M(j) \longleftarrow$ _____(3)

    else

        $M(j) \longleftarrow$ _____(4)

    end if

end for

4.7.2   *Context free grammars*

An *alphabet* is a finite set of symbols (also known as *terminals*), for example $\Sigma = \{0, 1\}$ or $\Sigma = \{a, b, c, \ldots, z\}$. Given an alphabet Σ, we denote the set of all finite strings over Σ as Σ^*. For example, if $\Sigma = \{0, 1\}$ then $\Sigma^* = \{\varepsilon, 0, 1, 00, 01, 10, 11, 000, 001, \ldots\}$, the set of all binary strings, where ε is the empty string. If w is a string, $|w|$ is the length of w, so $|\varepsilon| = 0$ and $|101001| = 6$.

A *context free grammar* is a tuple $G = (V, T, R, S)$, where V is a (finite) set of *variables*, T is a (finite) set of terminals, R is a (finite) set of rules, and S denotes the *starting variable*. All rules are of the form $X \longrightarrow \alpha$ where $X \in V$ and $\alpha \in (V \cup T)^*$, i.e., α is a (finite) string of variables and terminals.

A *derivation* of a string w is denoted as: $S \Rightarrow \alpha_1 \Rightarrow \alpha_2 \Rightarrow \cdots \Rightarrow \alpha_k$, where $\alpha_k = w$ and each α_{i+1} is obtained from α_i by one of the rules, meaning that a variable X in α_i is replaced by α to obtain α_{i+1} if $X \longrightarrow \alpha$ was one of the rules of the grammar.

The set $L(G)$, called *the language of the grammar* G, is the set of all those strings in T^* which have G-derivations.

Sometimes, for the sake of succinctness, if there are several rules for the same variable X, for example $X \longrightarrow \gamma_1, \cdots, X \longrightarrow \gamma_k$, we simply write $X \longrightarrow \gamma_1 | \cdots | \gamma_k$ as if it were one rule, but the symbol "|" denotes that this is a family of rules for X.

For example, consider $G_{\text{pal}} = (\{S\}, \{0,1\}, \{S \longrightarrow \varepsilon|0|1|0S0|1S1\}, S)$. In this grammar there is only one variable, S, which is also (necessarily) the starting variable, and there are five rules, which are represented in succinct form as a "single" rule $S \longrightarrow \varepsilon|0|1|0S0|1S1$, and there are two terminals $0, 1$. For example, $S \Rightarrow 0S0 \Rightarrow 01S10 \Rightarrow 01010$ shows that $01010 \in L(G_{\text{pal}})$. In fact, G_{pal} generates precisely the set of all palindromes[2] over 0,1, i.e., the set of all binary strings that read the same backwards as forwards.

Problem 4.22. *Show that $L_{pal} = L(G_{pal})$, in other words, show that G_{pal} is the grammar that generates precisely the set of all palindromes over $\{0,1\}$. To show "$\subseteq$" use induction on the length of strings to show that each palindrome has a derivation; for "$\supseteq$" use induction on the length of derivations.*

A context-free grammar is in *Chomsky Normal Form (CNF)* if all the rules are of the form: $A \longrightarrow BC$, $A \longrightarrow a$, and $S \longrightarrow \varepsilon$, where A, B, C are variables (not necessarily distinct), a is a terminal, and S is the starting variable (so note that ε can only be generated from the starting variable).

The grammar G_{pal} can be translated into CNF as follows: first the rule $S \longrightarrow \varepsilon|0|1$ is allowed as is, so we only have to deal with $S \longrightarrow 0S0$ and $S \longrightarrow 1S1$. To do that, we introduce two new variables U_0, U_1 and the following two rules, $U_0 \longrightarrow 0$ and $U_1 \longrightarrow 1$, and replace $S \longrightarrow 0S0|1S1$ with $S \longrightarrow U_0SU_0|U_1SU_1$. Now the only problem is that there are too many variables (we are only allowed 2 per body of rule in CNF), so we introduce two more new variables V_0, V_1 and transform $S \longrightarrow U_0SU_0|U_1SU_1$ as follows: $S \longrightarrow U_0V_0$, $V_0 \longrightarrow SU_0$, and $S \longrightarrow U_1V_1$, $V_1 \longrightarrow SU_1$. Putting it all together, we have a new grammar G'_{pal} given by $(\{S, U_0, U_1, V_0, V_1\}, \{0,1\}, R, S)$ and R is given by the following rules:

$$\{S \longrightarrow \varepsilon|0|1, U_0V_0, U_1V_1,\ U_0 \longrightarrow 0,\ U_1 \longrightarrow 1,\ V_0 \longrightarrow SU_0,\ V_1 \longrightarrow SU_1\},$$

---

[2]We already encountered palindromes on page 8, and we have examined the correctness of algorithm 1.3 for identifying palindromes.

and note that G'_{pal} is also a correct grammar for the language of palindromes, and furthermore it is in CNF.

Problem 4.23. *Every context-free grammar can be transformed to an equivalent grammar (i.e., generating the same set of strings of terminals) in CNF. (Hint: the main ideas are given in the above example; you will have to deal with the case $X \longrightarrow Y$ (the so called* unit *rules), and $X \longrightarrow \varepsilon$ where X is not the starting variable (the so called ε-rules).)*

Given a grammar G in CNF, and a string $w = a_1 a_2 \ldots a_n$, we can test whether $w \in L(G)$ using the CYK dynamic algorithm (algorithm 4.6). On input $G, w = a_1 a_2 \ldots a_n$ algorithm 4.6 builds an $n \times n$ table T, where each entry contains a subset of V. At the end, $w \in L(G)$ iff the start variable S is contained in position $(1, n)$ of T. The main idea is to put variable X_1 in position (i, j) if X_2 is in position (i, k) and X_3 is in position $(k + 1, j)$ and $X_1 \longrightarrow X_2 X_3$ is a rule. The reasoning is that X_1 is in position (i, k) iff $X_1 \overset{*}{\Rightarrow} a_i \ldots a_k$, that is, the substring $a_i \ldots a_k$ of the input string can be generated from X_1. Let $V = \{X_1, X_2, \ldots, X_m\}$.

Algorithm 4.6 CYK

---

    for $i = 1..n$ **do**

        for $j = 1..m$ **do**

            Place variable X_j in (i, i) iff $X_j \longrightarrow a_i$ is a rule of G

        end for

    end for

    for $1 \leq i < j \leq n$ **do**

        for $k = i..(j - 1)$ **do**

            if $(\exists X_p \in (i, k) \wedge \exists X_q \in (k + 1, j) \wedge \exists X_r \longrightarrow X_p X_q)$ **then**

                Put X_r in (i, j)

            end if

        end for

    end for

In the example in figure 4.5, we show which entries in the table we need to use to compute the contents of $(2, 5)$.

Problem 4.24. *Show the correctness of algorithm 4.6.*

x	(2,2)	(2,3)	(2,4)	**(2,5)**
x	x			(3,5)
x	x	x		(4,5)
x	x	x	x	(5,5)

Fig. 4.5   Computing the entry $(2, 5)$.

4.8   Answers to selected problems

Problem 4.4. Consider the graph G_n in figure 4.6. There are $2 + n + n = 2n + 2$ nodes in it, and there are 2^n paths from s to t. To see that, note that starting at s we have a choice to go to node 1 or node $1'$. And then we always have a choice to go up or down, so $2 \times 2^{n-1}$ paths that land us at n or n', and finally we just go to t. Note that we have given an undirected graph; but simply giving all the edges a "left-to-right" direction gives us an example for directed graphs.

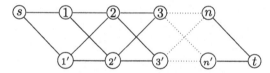

Fig. 4.6   Exponentially many paths (problem 4.4).

Problems 4.5 and 4.6. The precondition is that $\forall i, j \in [n]$ we have that $B(i, j) = A(0, i, j)$. The postcondition is that $\forall i, j \in [n]$ we have that $B(i, j) = A(n, i, j)$. The loop invariant is that after the k-th iteration of the main loop, $B(i, j) = A(k, i, j)$. To prove the loop invariant note that $B(i, j)$ is given by $\min\{B(i, j), B(i, k) + B(k, j)\}$, so the only worry is that $B(i, k)$ or $B(k, j)$ was *already* updated, so we are not getting $A(k - 1, i, k)$ or $A(k - 1, k, j)$ as we should, but rather $A(k, i, k)$ or $A(k, k, j)$. But, it turns out that $A(k, i, k) = A(k - 1, i, k)$ and $A(k, k, j) = A(k - 1, k, j)$, because the shortest path from i to k (or k to j) does *not* contain k as an intermediate node.

Problem 4.7. Overwriting does not create a problem, because the values of j are considered in decreasing order $C, C - 1, \ldots, 1$. Thus the array position $S(j - w_i)$ has not yet been updated when the reference is made.

Problem 4.8. The precondition is that for all j, $S(j) = R(0, j)$. The postcondition is that for all j, $S(j) = R(d, j)$. Let the loop invariant be the assertion that after the i-th step, $S(j) = R(i, j)$. This loop invariant holds since we start "filling" S in from the right, and we only change false to true (never true to false—the reason is that if we could build an intermediate value j with the first $(i - 1)$ weights, we can certainly still construct it with the first i weights).

Problem 4.9. Consider $w_1 = w_2 = 1$, $w_3 = 2$ and $C = 3$, so the table on this input would look as follows:

	0	1	2	3
	T	F	F	F
$w_1 = 1$	T	T	F	F
$w_2 = 1$	T	T	T	F
$w_3 = 2$	T	T	T	T

Now consider the row for $w_1 = 1$, and the entry for the column labeled with 2. That entry is an F, as it should be, but if the for-loop in algorithm 4.3 were not a decreasing loop, then we would update that entry to a T since for $j = 2$, we have that $2 \geq w_1$ and $S(2 - w_1) = $ T.

Problem 4.10. First, we need to find the solution; so we look in the last row (i.e., row d) for the largest non-zero j. That is, the solution is given by $M = \max_{0 \leq j \leq C}[R(d, j) = $ T$]$. Now we check if $R(d - 1, M) = $ T. If yes, then we know that weight w_d is not necessary, so we do not include it, and continue looking at $R(d - 2, M)$. If no, then because $R(d, M) = $ T, we know that $M - w_d \geq 0 \wedge R(d - 1, M - w_d) = $ T. So we include w_d, and continue looking at $R(d - 1, M - w_d)$. We stop when we reach the first column of the array.

Problem 4.11. The natural greedy algorithm that attempts to solve SKS is the following: order the weights from the heaviest to the lightest, and add them in that order for as long as possible. Assume that $\overline{M} \neq M$, and let S_0 be the result of this greedy procedure, i.e., a subset of $\{1, \ldots, d\}$ such that $K(S_0) = \overline{M}$. First show that there is at least one weight in S_0: If $S_0 = \emptyset$, then $\overline{M} = 0$, and all weights must be larger than C, but then $M = 0$, so $\overline{M} = M$ which is not the case by assumption. Now show that there is at least one weight not in S_0: if all the weights are in S, then again $\overline{M} = \sum_{i=1}^{d} w_i = M$. Finally, show now that $\overline{M} > \frac{1}{2}C$ by considering the first weight, call it w_j, which has been rejected after at least one weight has been added (note that such a weight must exist; we may assume that there are no weights larger than the capacity C, and if there are we can just not consider them; therefore, the first weight on the list is added, and then we know that some weight will come along which won't be added; we consider the first such weight): If $w_j \leq \frac{1}{2}C$, then the sum of the weights which are already in is $> \frac{1}{2}C$, so $\overline{M} > \frac{1}{2}C$. If $w_j > \frac{1}{2}C$, then, since the objects are ordered by greedy in non-increasing order of weights, the weights that are already in are $> \frac{1}{2}C$, so again $\overline{M} > \frac{1}{2}C$.

Problem 4.12. In the first space put $w_i + \sum_{j \in S} w_j \leq C$ and in the second space put $S \longleftarrow S \cup \{i\}$.

Problem 4.13. Define "S is promising" to mean that S can be extended, using weights which have *not* been considered yet, to an optimal solutions $S_{\max}$. At the end, when no more weights have been left to consider, the loop invariant still holds true, so S itself must be optimal.

We show that "S is promising" is a loop invariant by induction on the number of iterations. Basis case: $S = \emptyset$, so S is clearly promising. Induction step: Suppose that S is promising (so S can be extended, using weights which have not been considered yet, to $S_{\max}$). Let S' be S after one more iteration. Suppose $i \in S'$. Since $w_i \geq \sum_{j=i+1}^{d} w_j$, it follows that:

$$K(S') \geq K(S) + \sum_{j=i+1}^{d} w_j \geq K(S_{\max})$$

so S' is already maximal, and we are done. Suppose $i \notin S'$; then we have that $w_i + \sum_{j \in S} w_j > C$, so $i \notin S_{\max}$, so S' can be extended (using weights which have not been considered yet!) to $S_{\max}$. In either case, S' is promising.

Problem 4.14. $V(i,j) = 0$ if $i = 0$ or $j = 0$. And for $i, j > 0$, $V(i,j)$ is

$$\begin{cases} V(i-1,j) & \text{if } j < w_i \text{ or } R(i-1, j-w_i) = \mathsf{F} \\ \max\{v_i + V(i-1, j-w_i), V(i-1,j)\} & \text{otherwise} \end{cases}$$

To see that this works, suppose that $j < w_i$. Then weight i cannot be included, so $V(i,j) = V(i-1,j)$. If $R(i-1, j-w_i) = \mathsf{F}$, then there is no subset $S \subseteq \{1, \ldots, i\}$ such that $i \in S$ and $K(S) = j$, so again weight i is not included, and $V(i,j) = V(i-1,j)$.

Otherwise, if $j \geq w_i$ and $R(i-1, j-w_i) = \mathsf{T}$, then weight i may or may not be included in S. We take the case which offers more value: $\max\{v_i + V(i-1, j-w_i), V(i-1,j)\}$.

Problem 4.15. By changing the definition of $V(i,j)$ given in (4.2) to have $K(S) \leq j$ (instead of $K(S) = j$), we can take the recurrence given for V in the solution to problem 4.14 and simply get rid of the part "or

$R(i - 1, j - w_i) = \text{F}$" to obtain a recurrence for V that does not require computing R.

Problem 4.17. The algorithm must include a computation of the distinct finish times, i.e., the u_i's, as well as a computation of the array H. Here we just give the algorithm for computing A based on the recurrence (4.3). The assumption is that there are n activities and k distinct finish times.

Algorithm 4.7 Activity selection

$A(0) \longleftarrow 0$
for $j : 1..k$ **do**
    max $\longleftarrow 0$
    for $i = 1..n$ **do**
        if $f_i = u_j$ **then**
            if $p_i + A(H(i)) > $ max **then**
                max $\longleftarrow p_i + A(H(i))$
            end if
        end if
    end for
    if $A(j - 1) > $ max **then**
        max $\longleftarrow A(j - 1)$
    end if
    $A(j) \longleftarrow$ max
end for

Problem 4.18. We show how to find the actual set of activities: Suppose $k > 0$. If $A(k) = A(k - 1)$, then no activity has been scheduled to end at time u_k, so we proceed recursively to examine $A(k - 1)$. If, on the other hand, $A(k) \neq A(k - 1)$, then we know that some activity has been scheduled to end at time u_k. We have to find out which one it is. We know that in this case $A(k) = \max_{1 \leq i \leq n} \{p_i + A(H(i)) | f_i = u_k\}$, so we examine all activities i, $1 \leq i \leq n$, and output the (first) activity i_0 such that $A(k) = p_{i_0} + A(H(i_0))$ and $f_{i_0} \leq u_k$. Now we repeat the procedure with $A(H(i_0))$. We end when $k = 0$.

Problem 4.19. Initialization: $A(0, t) = 0$, $0 \leq t \leq t_n$. To compute $A(i, t)$ for $i > 0$ first define $t_{\min} = \min\{t, t_i\}$. Now

$$A(i, t) = \begin{cases} A(i - 1, t) & \text{if } t_{\min} < d_i \\ \max\{A(i - 1, t), p_i + A(i - 1, t_{\min} - d_i)\} & \text{otherwise} \end{cases}$$

Justification: If job i is scheduled in the optimal schedule, it finishes at time $t_{\min}$, and starts at time $t_{\min} - d_i$. If it is scheduled, the maximum possible profit is $A(i-1, t_{\min} - d_i) + p_i$. Otherwise, the maximum profit is $A(i-1, t)$.

Problem 4.20. $M = \max_{1 \leq j \leq n} M(j)$

Problem 4.21.

(1) $r_1 (= S_{11})$
(2) $M(j-1) > 0$
(3) $M(j-1) + r_j$
(4) r_j

4.9 Notes

Any algorithms textbook will have a section on dynamic programming; see for example chapter 15 in [Cormen *et al.* (2001)] and chapter 6 in [Kleinberg and Tardos (2006)].

Chapter 5

Online Algorithms

The algorithms that we have seen in the previous chapter were all *offline* algorithms, in the sense that the entire input was always presented. In this chapter we change our paradigm and consider *online* algorithms, where the input is presented piece by piece, in a serial fashion, and the algorithm has to make decisions based on incomplete information.

A typical example of an application is any caching discipline; consider a hard disk from which data is read into a random access memory. Typically, the random access memory is much smaller, and so it must be decided which data has to be overwritten with new data. New requests for data from the hard disk arrive continuously, and it is hard to predict future requests.

Thus we must overwrite parts of the random access memory with new requests, but we must do the overwriting judiciously, so that we minimize future *misses* (data that is required but not present in the random access memory, and so it has to be brought in from the hard disk). Minimizing the number of misses is difficult when the future requests are unknown.

Correctness in the context of online algorithm has a different nuance; it means that the algorithm minimizes strategic errors. That is, an online algorithm will do worse than a corresponding offline algorithm that sees the entire input, but we want it to be as competitive as possible—given its intrinsic limitations. Thus showing correctness will mean showing that the algorithm performs as well as can be expected given its limitations. Thus, the emphasis of correctness is on performance, which makes sense as online algorithms are used for such applications as caching (described in the above paragraph) that can be the performance bottleneck of the entire system.

We introduce online algorithms with the list accessing problem in section 5.1, and then present paging algorithms in section 5.2.

5.1   List accessing problem

We have a filing cabinet containing l labeled but unsorted files. We receive a sequence of requests to access files; each request is the label of a file. After a request to access file x is given, we must locate the file and remove it form the cabinet (to get some information from the file).

Suppose that the only way this can be done is by flipping through the files from the beginning until the file is located. In accessing the i-th file in the cabinet, we incur a cost (search time) of i; and if the file is not in the cabinet, we incur a cost of l (since we must scan the whole file cabinet). After taking out the file, we must return it to the cabinet before the next request for another file is made known; however, we may reorganize the cabinet, for instance, by reinserting the accessed file closer to the front of the cabinet. The incentive for such a reorganization is that it may save us some search time in the future; for example, if a certain file is requested frequently, it is wise to insert it closer to the front. Our goal is to find a reorganization rule that minimizes the search time.

Let $\sigma = \sigma_1, \sigma_2, \ldots, \sigma_n$ be a (finite) sequence of requests. To service request σ_i, a list accessing algorithm ALG must search for the requested item by traversing the list from the beginning, until it finds it. The cost of retrieving this item is the index of its position on the list. Thus, if item σ_i is in position j, the cost of retrieving it is j. Furthermore, the algorithm may reorganize the list at any time. The work associated with a reorganization is measured in terms of the minimum number of transpositions of consecutive items needed for this reorganization. However, immediately after accessing an item, we allow this item to be moved free of charge to any location closer to the front of this list.

Problem 5.1. *What is the justification for this "free move"? In other words, why does it make sense to allow placing an item "for free" right after accessing it?*

The transpositions associated with these "free moves" are called *free*, whereas all other transpositions are called *paid* and cost 1 each. Let $\text{ALG}(\sigma)$ be the sum of the costs of servicing all the items on the list σ, i.e., the sum of the costs of all the searches plus the sum of the costs of all paid transpositions.

For a given list accessing algorithm ALG, let $\text{ALG}_P(\sigma)$ be the number of paid transpositions made by *ALG* when processing the request sequence σ. Similarly, let $\text{ALG}_F(\sigma)$ be the number of free transpositions, and let

$\text{ALG}_C(\sigma)$ be the cost of all operations other than paid transpositions—for example the costs of the searches. Clearly, the total cost is given by $\text{ALG}(\sigma) = \text{ALG}_C(\sigma) + \text{ALG}_P(\sigma)$.

We consider first the *static list accessing model*, where we have a list of l items, and the only requests are to access an item on the list (i.e., there are no insertions or deletions). Many algorithms have been proposed for managing lists; we are going to examine *Move To Front (MTF)*, where after accessing an item, we move it to the front of the list, without changing the relative order of the other items.

Further, we assume that σ consists of *only* those items which appear on the list of MTF—this is not a crucial simplification; see problem 5.7. Notice that $\text{MTF}(\sigma) = \text{MTF}_C(\sigma)$, which is the sum of the costs of all the searches, since we only change the position of an item when we retrieve it (in which case we move it to the front).

Theorem 5.2. *Let* OPT *be an optimal (offline) algorithm for the static list accessing model. Suppose that* OPT *and* MTF *both start with the same list configuration. Then, for any sequence of requests σ, where $|\sigma| = n$, we have that* $\text{MTF}(\sigma) \leq 2 \cdot \text{OPT}_C(\sigma) + \text{OPT}_P(\sigma) - \text{OPT}_F(\sigma) - n$.

Proof. Imagine that both MTF and OPT process the requests in σ, while each algorithm works on its own list, starting from the same list. You may think of MTF and OPT as working in parallel, starting from the same list, and neither starts to process σ_i until the other is ready to do so.

Let

$$a_i = t_i + (\Phi_i - \Phi_{i-1}) \tag{5.1}$$

where t_i is the actual cost that MTF incurs for processing this request (so t_i is in effect the position of item σ_i on the list of MTF *after* the first $i - 1$ requests have been serviced). Φ_i is a *potential function*, and here it is defined as the number of *inversions* in MTF's list with respect to OPT's list. An inversion is defined to be an ordered pair of items x_j and x_k, where x_j precedes x_k in MTF's list, but x_k precedes x_j in OPT's list.

Problem 5.3. *Suppose that $l = 3$, and the list of* MTF *is x_1, x_2, x_3, and the list of* OPT *is x_3, x_2, x_1. What is Φ in this case? In fact, how can we compute* $\text{OPT}(\sigma)$, *where σ is an arbitrary sequence of requests, without knowing how* OPT *works?*

Note that Φ_0 depends only on the initial configurations of MTF and OPT, and since we assume that the lists are initially identical, $\Phi_0 = 0$.

Finally, the value a_i in (5.1) is called the *amortized cost*, and its intended meaning is the cost of accessing σ_i (i.e., t_i) plus a measure of the increase of the "distance" between MTF's list and OPT's list after processing σ_i (i.e., $\Phi_i - \Phi_{i-1}$).

It is obvious that the cost incurred by MTF in servicing σ, denoted MTF(σ), is $\sum_{i=1}^{n} t_i$. But instead of computing $\sum_{i=1}^{n} t_i$, which is difficult, we compute $\sum_{i=1}^{n} a_i$ which is much easier. The relationship between the two summations is,

$$\text{MTF}(\sigma) = \sum_{i=1}^{n} t_i = \Phi_0 - \Phi_n + \sum_{i=1}^{n} a_i, \tag{5.2}$$

and since we agreed that $\Phi_0 = 0$, and Φ_i is always positive, we have that,

$$\text{MTF}(\sigma) \leq \sum_{i=1}^{n} a_i. \tag{5.3}$$

So now it remains to compute an upper bound for a_i.

Problem 5.4. *Prove the second equality of equation (5.2).*

Assume that the i-th request, σ_i, is in position j of OPT, and in position k of MTF (i.e., this is the position of this item *after* the first $(i-1)$ requests have been completed). Let x_j denote this item—see figure 5.1.

We are going to show that

$$a_i \leq (2s_i - 1) + P_i - F_i, \tag{5.4}$$

where s_i is the search cost incurred by OPT for accessing request σ_i, and P_i and F_i are the paid and free transpositions, respectively, incurred by OPT when servicing σ_i. This shows that

$$\sum_{i=1}^{n} a_i \leq \sum_{i=1}^{n} ((2s_i - 1) + P_i - F_i)$$

$$= 2(\sum_{i=1}^{n} s_i) + (\sum_{i=1}^{n} P_i) - (\sum_{i=1}^{n} F_i) - n$$

$$= 2\text{OPT}_C(\sigma) + \text{OPT}_P(\sigma) - \text{OPT}_F(\sigma) - n,$$

which, together with the inequality (5.3), will show the theorem.

It remains to show (5.4), and we do so in two steps: in the first step MTF makes its move (i.e., moves x_j from the k-th slot to the beginning of its list) and we measure the change in the potential function *with respect to* the configuration of the list of OPT *before* OPT makes its own moves to deal with the request for x_j. In the second step, OPT makes its move and now we measure the change in the potential function *with respect to* the configuration of the list of MTF *after* MTF has completed its handling of the request (i.e., with x_j at the beginning of the list of MTF).

Fig. 5.1 x_j is in position k in MTF, and in position j in OPT. Note that in the figure it appears that $j < k$, but we make no such assumption in the analysis. Let * denote items located before x_j in MTF but after x_j in OPT, i.e., the * indicate inversions with respect to x_j. There may be other inversions involving x_j, namely items which are after x_j in MTF but before x_j in OPT, but we are not concerned with them.

See figure 5.1: suppose that there are v such *, i.e., v inversions of the type represented in the figure. Then, there are at least $(k - 1 - v)$ items that precede x_j in both list.

Problem 5.5. *Explain why at least $(k - 1 - v)$ items precede x_j in both lists.*

But this implies that $(k - 1 - v) \leq (j - 1)$, since x_j is in the j-th position in OPT. Thus,

$$(k - v) \leq j. \tag{5.5}$$

So what happens when MTF moves x_j to the front of the list? Two things happen (in terms of inversions): (i) $(k - 1 - v)$ new inversions are created, with respect to OPT's list, before OPT itself deals with the request for x_j. Also, (ii) v inversions are eliminated, again with respect to OPT's list, before OPT itself deals with the request for x_j.

Therefore, the contribution to the amortized cost is:

$$k + ((k - 1 - v) - v) = 2(k - v) - 1 \overset{(1)}{\leq} 2j - 1 \overset{(2)}{=} 2s - 1 \tag{5.6}$$

where (1) follows from (5.5), and (2) follows from the fact that the search cost incurred by OPT when looking for x_j is exactly j. Note that (5.6) looks similar to (5.4), but we are missing $+P_i - F_i$. These terms will come from considering the second step of the analysis: OPT makes its move and we measure the change of potential with respect to MTF with x_j at the beginning of the list. This is dealt with in the next problem.

Problem 5.6. *In the second step of the analysis,* MTF *has made its move and* OPT, *after retrieving* x_j, *rearranges its list. Show that each paid transposition contributes 1 to the amortized cost and each free transposition contributes* -1 *to the amortized cost.*

This finishes the proof.   □

Problem 5.7. *Consider the* dynamic list accessing model, *where besides accesses, we also have* insertions *(the cost of an insertion is* $l + 1$, *where* l *is the length of the list) and* deletions *(where the cost of a deletion is the same as the cost of an access, i.e., the position of the item on the list). Show that theorem 5.2 still holds for the dynamic case.*

An online algorithm is *c-competitive* if there is a constant α such that for all (finite) input sequences $\mathrm{ALG}(\sigma) \le c \cdot \mathrm{OPT}(\sigma) + \alpha$. The infimum over the set of all values c such that ALG is c-competitive is called the *competitive ratio* of ALG and is denoted $\mathcal{R}(\mathrm{ALG})$.

Problem 5.8. *Show that* $\mathrm{OPT}(\sigma) \le n \cdot l$, *where* l *is the length of the list and* n *is* $|\sigma|$.

Problem 5.9. *Show that* MTF *is a 2-competitive algorithm, and that* $\mathcal{R}(\mathrm{MTF}) \le 2 - \frac{1}{l}$.

The traditional approach to studying online algorithms falls within the framework of *distributional* (or *average-case*) complexity: a distribution on event sequences is hypothesizes, and the expected payoff per event is analyzed. However, we are interested in the (more recent) approach of *competitive analysis*, whereby the payoff of an online algorithm is measured by comparing its performance to that of an *optimal offline algorithm*. Competitive analysis thus falls within the framework of *worst case* complexity.

5.2 Paging

Consider the following two-level *virtual memory system*: each level (slow and fast) can store a number of fixed-size memory units called *pages*. The slow memory stores N pages, and the fast memory stores $k < N$ pages. The k is usually much smaller than N; this is sometimes denoted as $k \ll N$.

Given a request for page p_i, the system must make the page p_i available in the fast memory. If p_i is already in the fast memory (called a *hit*), the system need not do anything. Otherwise, on a *miss*, the system incurs one *page fault*, and must copy the page p_i from the slow memory to the fast memory. In doing so, the system is faced with the following problem: which page to evict from the fast memory to make space for p_i. In order to *minimize* the number of page faults, the choice of which page to evict must be made wisely.

Typical examples of pairs of fast and slow memory are the RAM and the hard disk, or the cache and the RAM. Often we will refer to the fast memory as the cache. Because of its important role in the performance of almost every computer system, the paging problem has been extensively studied since the 1960s.

A large variety of page replacement algorithms can be found in existing computer systems—see figure 5.2.

LRU	*Least Recently Used*
CLOCK	*Clock Replacement*
FIFO	*First-In/First-Out*
LIFO	*Last-In/First-Out*
LFU	*Least Frequently Used*
LFD	*Longest Forward Distance*

Fig. 5.2 Different page replacement algorithms. The last algorithm, LFD, is offline (as it has to see into the future to know the *longest forward distance*), while all the other algorithms are online.

All the caching disciplines in figure 5.2, except for the last one, are online algorithms; that is, they are algorithms that make decisions based on past events, rather than the future. The last algorithm, LFD, replaces the page whose next request is the latest, which requires full knowledge of future requests, and hence it is an offline algorithm. Furthermore, all the caching disciplines fall under the category of demand paging; they only evict on page faults. We take a close look at this scheme in the next section, and we shall see that this is not a restriction.

5.2.1 *Demand paging*

Demand paging algorithms never evict a page from the cache unless there is a page fault. We consider the *page fault model*: in this somewhat simplistic model, we charge 1 for bringing a page into the fast memory, and we charge 0 for accessing a page which is already there—in practice there are others costs involved. As the next theorem shows we lose nothing by restricting our attention to this model.

Theorem 5.10. *Any page replacement algorithm (online or offline) can be modified to be demand paging without increasing the overall cost on any request sequence.*

Proof. In a demand paging algorithm a page fault causes exactly one eviction (once the cache is full, that is), and there are no evictions between misses. So let ALG be any paging algorithm. We show how to modify it to make it a demand paging algorithm ALG′, in such a way that on any input sequence ALG′ incurs at most the cost (makes at most as many page moves from slow to fast memory) as ALG, i.e., $\forall \sigma$, $\text{ALG}'(\sigma) \leq \text{ALG}(\sigma)$.

Suppose that ALG has a cache of size k. Define ALG′ as follows: ALG′ also has a cache of size k, plus k registers. ALG′ runs a simulation of ALG, keeping in its k registers the page numbers of the pages that ALG would have in its cache. Based on the behavior of ALG, ALG′ makes decisions to evict pages.

Suppose page p is requested. If p is in the cache of ALG′, do nothing. Otherwise, if a page fault occurs, ALG′ behaves according to the following two cases:

Case 1. If ALG also has a page fault (that is, the number of p is *not* in the registers), and ALG evicts a page from register i to make room for p, then ALG′ evicts a page from slot i in its cache, to make room for p.

Case 2. If ALG does not have a page fault, then the number of p must be in, say, register i. In that case, ALG′ evicts the contents of slot i in its cache, and moves p in there.

Thus ALG′ is a demand paging algorithm.

We now show that ALG′ incurs at most the cost of ALG on any input sequence; that is, ALG′ has at most as many page faults as ALG. To do this, we pair each page move of ALG′ with a page move of ALG in a unique manner as follows: If ALG′ and ALG both incur a page fault, then match the corresponding page moves. Otherwise, if ALG already had the page in its cache, it must have moved it there before, so match that move with the current move of ALG′.

It is never the case that two different moves of ALG′ are matched with a single move of ALG. To see this, suppose that on some input sequence, we encounter for the first time the situation where two moves of ALG′ are matched with the same move of ALG. This can only happen in the following situation: page p is requested, ALG′ incurs a page fault, it moves p into its cache, and we match this move with a past move of ALG, which has been matched already! But this means that page p was already requested, and after it has been requested, it has been evicted from the cache of ALG′ (otherwise, ALG′ wouldn't have had a page fault). Thus, ALG′ evicted page p while ALG did not, so they were not in the same slot. But ALG′ put (the first time) p in the same slot as ALG! Contradiction. Therefore, we couldn't have matched a move twice. Thus, we can match each move of ALG′ with a move of ALG, in a one-to-one manner, and hence ALG′ makes at most as many moves as ALG. See figure 5.3.

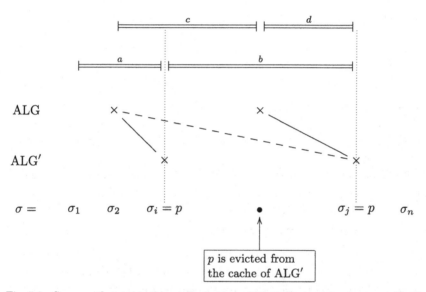

Fig. 5.3 Suppose that i, j is the smallest pair such that there exists a page p with the property that $\sigma_i = \sigma_j = p$ and ALG′ incurs a page fault at σ_i and σ_j and the two corresponding page moves of ALG′ are both matched with the same page move of p by ALG somewhere in the stretch a. We show that this is not possible: if ALG′ incurs a page fault at $\sigma_i = \sigma_j = p$ it means that somewhere in b the page p is evicted—this point is denoted with '•'. If ALG did not evict p in the stretch c, then ALG also evicts page p at '•' and so it must then bring it back to the cache in stretch d—we would match the $\times$ at σ_j with that move. If ALG did evict p in the stretch c, then again it would have to bring it back in before σ_j. In any case, there is a later move of p that would be matched with the page fault of ALG′ at σ_j.

Problem 5.11. *In figure 5.3 we postulate the existence of a "smallest" pair i, j with the given properties. Show that if such a pair exists then there* *exists a "smallest" such pair; what does "smallest" mean in this case?*

The idea is that ALG does not gain anything by moving a page into its cache preemptively (before the page is actually needed). ALG′ waits for the request before taking the same action.

In the meantime (between the time that ALG moves in the page and the time that it is requested and ALG′ brings it in), ALG′ can only gain, because there are no requests for that page during that time, but there might be a request for the page that ALG evicted preemptively.

Note that in the simulation, ALG′ only needs k extra registers, to keep track of the page numbers of the pages in the cache of ALG, so it is an *efficient* simulation.   □

5.2.2   *FIFO*

When a page must be replaced, the *oldest* page is chosen. It is not necessary to record the time when a page was brought in; all we need to do is create a FIFO (First-In/First-Out) queue to hold all pages in memory. The FIFO algorithm is easy to understand and program, but its performance is not always very good.

FIFO also suffers from the so called *Belady's anomaly*. Suppose that we have the following sequence of page requests: $1, 2, 3, 4, 1, 2, 5, 1, 2, 3, 4, 5$. Then, we have more page faults when $k = 4$ than when $k = 3$. That is, FIFO has more page faults with a bigger cache!

Problem 5.12. *For a general i, provide a sequence of page requests that illustrates Belady's anomaly incurred by FIFO on cache sizes i and $i + 1$. In your analysis, assume that the cache is initially empty.*

5.2.3   *LRU*

The optimal algorithm for page replacement, OPT, evicts the page whose next request is the latest (and if some pages are never requested again, then one of them must be evicted, and the choice of which one is unimportant). This is an impractical algorithm from the point of view of online algorithms as we do not know the future.

However, if we use the recent past as an approximation of the near future, then we will replace the page that *has not been used for the longest period of time*. This approach is the *Least Recently Used (LRU)* algorithm.

LRU replacement associates with each page the time of that page's last use. When a page must be replaced, LRU chooses that page that has not been used for the longest period of time. The LRU algorithm is considered to be good, and is often implemented—the major problem is *how* to implement it; two typical solutions are counters and stacks.

Counters: Keep track of the time when a given page was last referenced, updating the counter every time we request it. This scheme requires a search of the page table to find the LRU page, and a write to memory for each request; an obvious problem might be clock overflow.

Stack: Keep a stack of page numbers. Whenever a page is referenced, it is removed from the stack and put on the top. In this way, the top of the stack is always the most recently used page, and the bottom is the LRU page. Because entries are removed from the middle of the stack, it is best implemented by a doubly-linked list.

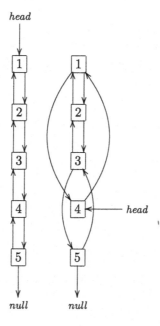

Fig. 5.4 LRU stack implementation with a doubly-linked list. The requested page is page 4; the left list shows the state *before* page 4 is requested, and the right list shows the state *after* the request has been serviced.

How many pointer operations need to be performed in the example in figure 5.4? Six, if we count as follows: remove old head and add new head (2 operations), connect 4 with 1 (2 operations), connect 3 with 5 (2 operations). However, we could have also counted disconnecting 3 with 4 and 4 with 5, giving 4 more pointer operations, giving us a total of 10. A third strategy would be not to count disconnecting pointers, in which case we would get half of these operations, 5. It does not really matter how we count, because the point is that in order to move a requested page (after a hit) to the top, we require a small *constant* number of pointer operations, regardless of how we count them.

Problem 5.13. *List the pointer operations that have to be performed if the requested page is not in the cache. Note that you should list the pointer operations (not just give a "magic number"), since we just showed that there are three different (all reasonable) ways to count them. Again, the point is, that if a page has to be brought in from the slow memory to the cache, a small constant number of pointer operations have to be performed.*

Problem 5.14. *We have implemented LRU with a doubly-linked list. What would be the problem if we used a normal linked list instead? That is, if every page had only a pointer to the next page: $i \rightsquigarrow j$, meaning that i was requested more recently than j, but no page was requested later than i and sooner than j.*

Lemma 5.15. LRU *does not incur Belady's anomaly (on any cache size and any request sequence).*

Proof. Let σ be any request sequence, $\sigma = p_1, p_2, \ldots, p_n$, and, as usual, let $\mathrm{LRU}_i(\sigma)$ be the number of faults that LRU incurs on σ with a cache of size i. We are going to show that for all i and σ, the following property holds:

$$\mathrm{LRU}_i(\sigma) \geq \mathrm{LRU}_{i+1}(\sigma). \tag{5.7}$$

Once we show (5.7), it follows that for any pair $i < j$ and any request sequence σ, $\mathrm{LRU}_i(\sigma) \geq \mathrm{LRU}_j(\sigma)$, and conclude that LRU does not incur Belady's anomaly.

To show (5.7), we define a property of doubly-linked lists which we call "embedding." We say that a doubly-linked list of size i can be *embedded* in another doubly-linked list of size $i + 1$, if the two doubly-linked lists are identical, except that the longer one may have one more item at the "end." See figure 5.5, where the doubly-linked list of size 3 on the left can be embedded in the doubly-linked list of size 4 on the right.

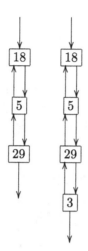

Fig. 5.5   The list on the left can be *embedded* into the list on the right.

At the beginning of processing the request sequence, when the caches are getting filled up, the two lists are identical, but once the caches are full, the LRU_{i+1} cache will have one more item.

Claim 5.16. *After each step of processing a sequence of requests, the doubly-linked list of* LRU_i *can be embedded into the doubly-linked list of* LRU_{i+1}.

Proof.   We prove this claim by induction on the number of steps. Basis case: if $n = 1$, then both LRU_i and LRU_{i+1} incur a fault and bring in p_1. Induction step: suppose that the claim holds after step n; we show that it also holds after step $n + 1$. Consider the following cases: (1) LRU_i has a hit on p_{n+1}, (2) LRU_i has a fault on p_{n+1}, (2a) LRU_{i+1} also has a fault, (2b) LRU_{i+1} does not have a fault.

Problem 5.17. *Show that in each case the embedding property is being preserved.*

This finishes the proof of the claim.   □

Problem 5.18. *Use the claim to prove (5.7).*

This finishes the proof of the lemma.   □

5.2.4 *Marking algorithms*

Consider a cache of size k and fix a request sequence σ. We divide the request sequence into *phases* as follows: phase 0 is the empty sequence. For every $i \geq 1$, phase i is the maximal sequence following phase $i - 1$ that contains at most k distinct page requests; that is, if it exists, phase $i + 1$ begins on the request that constitutes the $k + 1$ distinct page request since the start of the i-th phase. Such a partition is called a k-*phase partition*. This partition is well defined and is independent of any particular algorithm processing σ.

For example, a 3-phase partition:

$$\underbrace{1,2,1,2,1,2,3}_{\text{3-phase \#1}},\underbrace{4,5,6,6,6,6,6,6,6,4,5,4}_{\text{3-phase \#2}},\underbrace{7,7,7,7,1,2}_{\text{3-phase \#3}}.$$

Let σ be any request sequence and consider its k-phase partition. Associate with each page a bit called the *mark*. The marking is done for the sake of analysis (this is not implemented by the algorithm, but "by us" to keep track of the doings of the algorithm). For each page, when its mark bit is set we say that the page is *marked*, and otherwise, *unmarked*.

Suppose that at the beginning of each k-phase we unmark all the pages, and we mark a page when it is first requested during the k-phase. A *marking algorithm* never evicts a marked page from its fast memory.

For example, suppose that $k = 2$, and σ is a request sequence. We show the 2-phases of σ:

$$\sigma = \underbrace{1,1,3,1}_{\text{2-phase \#1}},\underbrace{5,1,5,1,5,1}_{\text{2-phase \#2}},\underbrace{3,4,4,4}_{\text{2-phase \#3}},\underbrace{2,2,2,2}_{\text{2-phase \#4}}. \tag{5.8}$$

See figure 5.6 to examine the marking in this example. Note that after each phase, every page is unmarked and we begin marking afresh, and except for the last phase, all phases are always complete (they have exactly k distinct requests, 2 in this case).

With a marking algorithm, once a request for page p in phase i is made, p stays in the cache until the end of phase i—the first time p is requested, it is marked, and it stays marked for the entire phase, and a marking algorithm never evicts a marked page.

The intuition is that marking algorithms are good schemes for page replacement because, in any given phase, there are at most k distinct pages, so they all fit in a cache of size k, so it does not make sense to evict them in that phase (we can only lose by evicting, as the evicted page might be requested again).

step	1	2	3	4	5	step	1	2	3	4	5
1	x					10	x				x
2	x					11		x			
3	x		x			12		x	x		
4	x		x			13		x	x		
5					x	14		x	x		
6	x				x	15	x				
7	x				x	16	x				
8	x				x	17	x				
9	x				x	18	x				

Fig. 5.6   Marking in example (5.8).

Theorem 5.19. LRU *is a marking algorithm*

Proof.   We argue by contradiction; suppose that LRU on a cache of size k is not a marking algorithm. Let σ be a request sequence where there exists a k-phase partition, during which some marked page p is evicted. Consider the first request for p during this k-phase:

$$\sigma = p_1, p_2, p_3, \ldots, \ldots, \underbrace{\ldots, p, \ldots, \ldots, \ldots,}_{k\text{-phase}} \ldots, \ldots$$

Immediately after p is serviced, it is marked as the most recently used page in the cache (i.e., it is put at the top of the doubly-linked list).

In order for p to leave the cache, LRU must incur a page fault while p is the least recently used page. It follows that during the k-phase in question, $k+1$ distinct pages were requested: there are the $k-1$ pages that pushed p to the end of the list, there is p, and the page that got p evicted. Contradiction; a k-phase has at most k distinct pages. □

5.2.5   *FWF*

Flush When Full (FWF) is a very naïve page replacement algorithm that works as follows: whenever there is a page fault and there is no space left in the cache, evict all pages currently in the cache—call this action a "flush."

More precisely, we consider the following version of the FWF algorithm: each slot in the cache has a single bit associated with it. At the beginning, all these bits are set to zero. When a page p is requested, FWF checks only the slots with a marked bit. If p is found, it is serviced. If p is not found, then it has to be brought in from the slow memory (even if it actually is in the cache, in an unmarked slot). FWF looks for a slot with a zero bit, and one of the following happens: (1) a slot with a zero bit (an unmarked page) is found, in which case FWF replaces that page with p. (2) a slot with a zero bit is not found (all pages are marked), in which case FWF unmarks all the slots, and replaces any page with p, and it marks p's bit.

Problem 5.20. *Show that* FWF *is a marking algorithm. Show that* FIFO *is not a marking algorithm.*

Problem 5.21. *A page replacement algorithm* ALG *is conservative if, on any consecutive input subsequence containing* k *or fewer distinct page requests,* ALG *will incur* k *or fewer page faults. Prove that* LRU *and* FIFO *are conservative, but* FWF *is not.*

5.2.6   *LFD*

The optimal page replacement algorithm turns out to be LFD (*Longest Forward Distance*—see figure 5.2). LFD evicts the page that will not be used for the longest period of time, and as such, it cannot be implemented in practice because it requires knowledge of the future. However, it is very useful for comparison studies, i.e., competitive analysis.

Theorem 5.22. LFD *is the optimal (offline) page replacement algorithm, i.e.,* OPT = LFD.

Proof.   We will show that if ALG is any paging algorithm (online or offline), then on any sequence of requests σ, $\text{ALG}(\sigma) \geq \text{LFD}(\sigma)$. As usual, $\text{ALG}(\sigma)$ denotes the number of page faults of ALG on the sequence of requests σ. We assume throughout that all algorithms are working with a cache of a fixed size k. We need to prove the following claim.

Claim 5.23. *Let* ALG *be any paging algorithm. Let* $\sigma = p_1, p_2, \ldots, p_n$ *be any request sequence. Then, it is possible to construct an offline algorithm* ALG_i *that satisfies the following three properties:*

(1) ALG_i *processes the first* $i - 1$ *requests of* σ *exactly as* ALG *does,*

(2) if the i-*th request results in a page fault,* ALG_i *evicts from the cache the page with the "longest forward distance,"*

(3) $\text{ALG}_i(\sigma) \leq \text{ALG}(\sigma)$

Proof. Divide σ into three segments as follows:

$$\sigma = \sigma_1, p_i, \sigma_2,$$

where σ_1 and σ_2 each denote a block of requests.

Recall the proof of theorem 5.10 where we simulated ALG with ALG′ by running a "ghost simulation" of the contents of the cache of ALG on a set of registers, so ALG′ would know what to do with its cache based on the contents of those registers. We do the same thing here: ALG_i runs a simulation of ALG on a set of registers.

As on σ_1, ALG_i is just ALG, it follows that $\text{ALG}_i(\sigma_1) = \text{ALG}(\sigma_1)$, and also, they both do or do not incur a page fault on p_i. If they do not, then let ALG_i continue behaving just like ALG on σ_2, so that $\text{ALG}_i(\sigma) = \text{ALG}(\sigma)$.

However, if they do incur a page fault on p_i, ALG_i evicts the page with the longest forward distance from its cache, and replaces it with p_i. If ALG also evicts the same page, then again, let ALG_i behave just like ALG for the rest of σ, so that $\text{ALG}_i(\sigma) = \text{ALG}(\sigma)$.

Finally, suppose that they both incur a fault at p_i, but ALG evicts some page q and ALG_i evicts some page p, and $p \neq q$; see figure 5.7. If both $p, q \notin \sigma_2$, then let ALG_i behave just like ALG, except the slots with p and q are interchanged (that is, when ALG evicts from the q-slot, ALG_i evicts from the p-slot, and when ALG evicts from the p-slot, ALG_i evicts from the q-slot).

ALG:		$\cancel{q}$			p	
ALG_i:		q			$\cancel{p}$	

Fig. 5.7 ALG evicts q and ALG_i evicts p, denoted with $\cancel{q}$ and $\cancel{p}$, respectively, and they both replace their evicted page with p_i.

If $q \in \sigma_2$ but $p \notin \sigma_2$, then again let ALG_i, when forced with an eviction, act just like ALG with the two slots interchanged. Note that in this case it may happen that $\text{ALG}_i(\sigma_2) < \text{ALG}(\sigma_2)$, since ALG evicted q, which is going to be requested again, but ALG_i evicted p which will never be requested.

Problem 5.24. *Why the case $q \notin \sigma_2$ and $p \in \sigma_2$ is not possible?*

Otherwise, we can assume that ALG_i evicts page p and ALG evicts page q, $p \neq q$, and:

$$\sigma_2 = p_{i+1}, \ldots, q, \ldots, p, \ldots, p_n. \tag{5.9}$$

Assume that the q shown in (5.9) is the earliest instance of q in σ_2. As before, let ALG_i act just like ALG with the q-slot and p-slot interchanged. We know for sure that ALG will have a fault at q. Suppose ALG does not have a fault at p; then, ALG never evicted p, so ALG_i never evicted q, so ALG_i did not have a fault at q. Therefore, $\text{ALG}_i(\sigma_2) \leq \text{ALG}(\sigma_2)$. $\qquad\square$

We now show how to use claim 5.23 to prove that LFD is in fact the optimal algorithm. Let $\sigma = p_1, p_2, \ldots, p_n$ be any sequence of requests. By the claim, we know that: $\text{ALG}_1(\sigma) \leq \text{ALG}(\sigma)$. Applying the claim again, we get $(\text{ALG}_1)_2(\sigma) \leq \text{ALG}_1(\sigma)$. Define $\overline{\text{ALG}}_j$ to be $(\cdots((\text{ALG}_1)_2)\cdots)_j$. Then, we obtain that $\overline{\text{ALG}}_j(\sigma) \leq \overline{\text{ALG}}_{j-1}(\sigma)$.

Note that $\overline{\text{ALG}}_n$ acts just like LFD on σ, and therefore we have that $\text{LFD}(\sigma) = \overline{\text{ALG}}_n(\sigma) \leq \text{ALG}(\sigma)$, and we are done. $\qquad\square$

Henceforth, OPT can be taken to be synonymous with LFD.

Theorem 5.25. *Any marking algorithm* ALG *is* $\left(\frac{k}{k-h+1}\right)$-*competitive, where k is the size of its cache, and h is the size of the cache of OPT.*

Proof. Fix any request sequence σ and consider its k-phase partition. Assume, for now, that the last phase of σ is complete (in general, the last phase may be incomplete).

Claim 5.26. *For any phase $i \geq 1$, a marking algorithm* ALG *incurs at most k page faults.*

Proof. This follows because there are k distinct page references in each phase. Once a page is requested, it is marked and therefore cannot be evicted until the phase has been completed. Consequently, ALG cannot fault twice on the same page. $\qquad\square$

If we denote the i-th k-phase of σ by σ_i, we can express the above claim as $\text{ALG}(\sigma_i) \leq k$. Thus, if there are s phases, $\text{ALG}(\sigma) \leq s \cdot k$.

Claim 5.27. $\text{OPT}(\sigma) \geq s \cdot (k - h + 1)$, *where again we assume that the requests are $\sigma = \sigma_1, \sigma_2, \ldots, \sigma_s$, where σ_s is complete.*

Proof. Let p_a be the first request of phase i, and p_b the last request of phase i. Suppose first that phase $i + 1$ exists (that is, i is *not* the last phase). Then, we partition σ into k-phases (even though the cache of OPT is of size k, we still partition σ into k-phases):

$$\sigma = \ldots, p_{a-1}, \underbrace{p_a, p_{a+1}, \ldots, p_b}_{k\text{-phase }\#i}, \underbrace{p_{b+1}, \ldots, \ldots}_{k\text{-phase }\#i+1}$$

After processing request p_a, OPT has at most $h - 1$ pages in its cache, not including p_a. From (and including) p_{a+1} until (and including) p_{b+1}, there are at least k distinct requests. Therefore, OPT must incur at least $k - (h - 1) = k - h + 1$ faults on this segment. To see this, note that there are two cases.

Case 1. p_a appears again in $p_{a+1}, \ldots, p_{b+1}$; then there are at least $(k + 1)$ distinct requests in the segment $p_{a+1}, \ldots, p_{b+1}$, and since OPT has a cache of size h, regardless of the contents of the cache, there will be at least $(k + 1) - h = k - h + 1$ page faults.

Case 2. Suppose that p_a does *not* appear again in $p_{a+1}, \ldots, p_{b+1}$, then since p_a is requested at the beginning of phase i, it is for sure in the cache by the time we start servicing $p_{a+1}, \ldots, p_{b+1}$. Since it is not requested again, it is taking up a spot in the cache, so at most $(h - 1)$ slots in the cache can be taken up by some of the elements requested in $p_{a+1}, \ldots, p_{b+1}$; so again, we have at least $k - (h - 1) = k - h + 1$ many faults.

If i is the last phase (so $i = s$), we do not have p_{b+1}, so we can only say that we have at least $k - h$ faults, but we make it up with p_1 which has not been counted. $\qquad\square$

It follows from claims 5.26 and 5.27 that:

$$\text{ALG}(\sigma) \leq s \cdot k \quad \text{and} \quad \text{OPT}(\sigma) \geq s \cdot (k - h + 1),$$

so that:

$$\frac{\text{ALG}(\sigma)}{s \cdot k} \leq 1 \leq \frac{\text{OPT}(\sigma)}{s \cdot (k - h + 1)},$$

so finally:

$$\text{ALG}(\sigma) \leq \left(\frac{k}{k - h + 1} \right) \cdot \text{OPT}(\sigma).$$

In the case that σ can be divided into s complete phases.

As was mentioned above, in general, the last phase may not be complete. Then, we repeat this analysis with $\sigma = \sigma_1, \sigma_2, \ldots, \sigma_{s-1}$, and for σ_s we use α at the end, so we get:

$$\text{ALG}(\sigma) \leq \left(\frac{k}{k - h + 1} \right) \cdot \text{OPT}(\sigma) + \alpha$$

Problem 5.28. *Work this out.*

Therefore, in either case we obtain that any marking algorithm ALG is $\left(\frac{k}{k - h + 1} \right)$-competitive. $\qquad\square$

5.3 Answers to selected problems

Problem 5.1. Think of the filing cabinet mentioned at the beginning of this chapter. As we scan the filing cabinet while searching for a particular file, we keep a pointer at any location along the way and then insert the accessed file in that location almost free of additional search or reorganization costs. We also assume that it wouldn't make sense to move the file to a later location.

Problem 5.3. The answer is 3. Note that in a list of n items there are $\binom{n}{2} = \frac{n \cdot (n-1)}{2}$ *un*ordered pairs (and $n \cdot (n-1)$ ordered pairs), so to compute Φ, we enumerate all those pairs, and increase a counter by 1 (starting from 0) each time we encounter an inversion. For the second question, note that while we do not know how OPT works exactly, we know that it services σ with the optimal cost, i.e., it services σ in the cheapest way possible. Thus, we can find OPT(σ) by an exhaustive enumeration: given our list $x_1, x_2, \ldots, x_l$ and a sequence of requests $\sigma = \sigma_1, \sigma_2, \ldots, \sigma_n$, we build a tree where the root is labeled with $x_1, x_2, \ldots, x_l$, and the children of the root are all the $l!$ permutations of the list. Then each node in turn has $l!$ many children; the depth of the tree is n. We calculate the cost of each branch and label the leaves with those costs. The cost of each branch is the sum of the costs of all the transpositions required to produce each consecutive node, and the costs of the searches associated with the corresponding list configurations. The cheapest branch (and there may be several) is precisely OPT(σ).

Problem 5.5. The number of elements before x_j in MTF is $(k-1)$, since x_j is in the k-th position. Of these $(k-1)$ elements, v are $*$. Both lists contain exactly the same elements, and the non-$*$ before x_j in MTF, $(k-1-v)$ must all be before x_j in OPT (if an element is before x_j in MTF and after x_j in OPT, then by definition it would be a $*$).

Problem 5.6. In the case of a paid transposition, the only change in the number of inversions can come from the two transposed items, as the relative order with respect to all the other items remains the same. In the case of a free transposition, we know that MTF already put the transposed item x_j at the front of its list, and we know that free transpositions can only move x_j forward, so the number of items before x_j in OPT decreases by 1.

Problem 5.8. OPT is the optimal (offline) algorithm, and hence it must do better than any other algorithm ALG. Thus, if we present an algorithm ALG such that for any σ we have ALG$(\sigma) \leq n \cdot l$ we would be done.

Here is a candidate for ALG: service all requests one-by-one without ever reorganizing the list. For each of the n requests the worst that can happen is that it is at the end of the list, i.e., in position l.

Problem 5.9. By theorem 5.2 we know that

$$\text{MTF}(\sigma) \leq 2 \cdot \text{OPT}_C(\sigma) + \text{OPT}_P(\sigma) - \text{OPT}_F(\sigma) - n$$

and the RHS is

$$\leq 2 \cdot \text{OPT}_C(\sigma) + \text{OPT}_P(\sigma) \leq 2 \cdot (\text{OPT}_C(\sigma) + \text{OPT}_P(\sigma)) = 2 \cdot \text{OPT}(\sigma).$$

This shows that MTF is 2-competitive (with $\alpha = 0$). For the second part, we repeat the above argument, but without "losing" the n factor, so we have $\text{MTF}(\sigma) \leq 2 \cdot \text{OPT}(\sigma) - n$. On the other hand, $\text{OPT}(\sigma) \leq n \cdot l$ (by problem 5.8), so

$$2 \cdot \text{OPT}(\sigma) - n \leq \left(2 - \frac{1}{l}\right) \cdot \text{OPT}(\sigma)$$

Problem 5.11. In the proof of theorem 5.10 we define a matching between the page moves (from slow memory into the cache) of ALG and ALG$'$. In order to show that the matching is one-to-one we postulate the existence of a pair i, j, $i \neq j$, with the following properties: (i) there exists a page p such that $\sigma_i = \sigma_j = p$, (ii) ALG$'$ incurs a page fault at σ_i and σ_j, and (iii) ALG$'$ has to move p into the cache to service σ_i and σ_j and those two moves are matched with the same move of p by ALG. For the sake of the argument in the proof of theorem 5.10 we want the "smallest" such pair—so we use the Least Number Principle (see page 3) to show that if such pairs exist at all, there must exist pairs where $i + j$ is minimal; we take any such pair.

Problem 5.12. Consider the following list:

$$\underbrace{1, 2, 3 \ldots, i, i+1}_{1}, \underbrace{1, 2, 3, \ldots, i-1}_{2}, \underbrace{i+2}_{3}, \underbrace{1, 2, 3, \ldots, i, i+1, i+2}_{4}.$$

If we have a cache of size $i + 1$, then we incur $i + 1$ faults in segment 1 (because the cache is initially empty), then we have $i - 1$ hits in segment 2, then we have another page fault in segment 3 so we evict 1, and in segment 4 we lag behind by 1 all the way, so we incur $i + 2$ page faults. Hence, we incur $i + 1 + 1 + i + 2 = 2i + 4$ page faults in total.

Suppose now that we have a cache of size i. Then we incur $i + 1$ page faults in segment 1, then we have $i - 1$ page faults in segment 2, and one page fault in segment 3, hence $2i + 1$ page faults before starting segment 4. When segment 4 starts, we already have pages 1 through $i - 1$ in the cache, so we have hits, and then when $i + 1$ is requested, we have a fault, and when $i + 2$ is requested we have a hit, and hence only one fault in segment 4. Therefore, we have $2i + 2$ page faults with a cache of size i. To understand this solution, make sure that you keep track of the contents of the cache after each of the four segments has been processed. Note that i has to be at least 3 for this example to work.

Problem 5.14. The problem with a singly-linked list is that to find the predecessor of a page we need to start the search always at the beginning of the list, increasing the overhead of maintaining the stack.

Problem 5.20. FWF really implements the marking bit, so it is almost a marking algorithm by definition. FIFO is not a marking algorithm because with $k = 3$, and the request sequence $1, 2, 3, 4, 2, 1$ it will evict 2 in the second phase even though it is marked.

Problem 5.21. We must assume that the cache is of size k. Otherwise the claim is not true: for example, suppose that we have a cache of size 1, and the following sequences: $1, 2, 1, 2$. Then, in that sequence of 4 requests there are only 2 distinct requests, yet with a cache of size 1, there would be 4 faults, for *any* page-replacement algorithm. With a cache of size k, LRU is never going to evict a page during this consecutive subsequence, once the page has been requested. Thus, each distinct page request can only cause one fault. Same goes for FIFO. Thus, they are both conservative algorithms. However, it is possible that half-way through the consecutive subsequence, the cache of FWF is going to get full, and FWF is going to evict everybody. Hence, FWF may have more than one page fault on the same page during this consecutive subsequence.

Problem 5.24. If $p \in \sigma_2$ and $q \notin \sigma_2$, then q would have a "longer forward distance" than p, and so p would not have been evicted by ALG. Rather, ALG would have evicted q or some other page that was not to be requested again.

5.4   Notes

An exhaustive text book on online algorithms is [Borodin and El-Yaniv (1998)]. Indeed, the material in this chapter is based on the presentation in that book; in particular, theorem 5.2, Fig. 5.1 and Problem 5.9 in this book are theorem 1.1, Fig. 1.1 and Exercise 1.3, respectively, in [Borodin and El-Yaniv (1998)], with the permission of the book's copyright owner, Cambridge University Press.

Chapter 6

Randomized Algorithms

It is very interesting that we can design procedures which, when confronted with a profusion of choices, instead of laboriously examining all the possible answers to those choices, they flip a coin to decide which way to go, and still "tend to" obtain the right output.

Obviously we save time when we resort to randomness, but what is very surprising is that the output of such procedures can be meaningful. That is, there are problems that computationally appear very difficult to solve, but when allowed the use of randomness it is possible to design procedures that solve those hard problems in a satisfactory manner: the output of the procedure is correct with a small probability of error. In fact this error can be made so small that it becomes negligible (say 1 in 2^{100}—the estimated number of atoms in the observable universe). Thus, many experts believe that the definition of "feasibly computable" ought to be computable in polynomial time with randomness, rather than just in polynomial time.

The advent of randomized algorithms came with the problem of primality testing, which in turn was spurred by the then burgeoning field of cryptography. Historically the first such algorithm was due to [Solovay and Strassen (1977)]. Primality testing remains one of the best problems to showcase the power of randomized algorithms; in this chapter we present the Rabin-Miller algorithm that came after the Solovay-Strassen algorithm, but it is somewhat simpler. Interestingly enough, there is now known a polynomial time algorithm for primality testing (due to [Agrawal *et al.* (2004)]), but the Rabin-Miller algorithm is still used in practice as it is more efficient (and its probability of error is negligible).

In this chapter we present three examples of randomized algorithms: an algorithm for perfect matching, for string pattern matching and finally the Rabin-Miller algorithm. We close with a discussion of cryptography.

6.1 Perfect matching

Consider a bipartite graph $G = (V \cup V', E)$, where $E \subseteq V \times V'$, and its adjacency matrix is defined as follows: $(A_G)_{ij} = x_{ij}$ if $(i, j') \in E_G$, and $(A_G)_{ij} = 0$ otherwise. See the example given in figure 6.1.

$$\begin{pmatrix} x_{11} & 0 & 0 & 0 \\ 0 & x_{22} & 0 & 0 \\ 0 & 0 & 0 & x_{34} \\ 0 & 0 & x_{43} & 0 \end{pmatrix}$$

Fig. 6.1   On the left we have a bipartite graph $G = (V \cup V', E)$ where $V = \{1, 2, 3, 4\}$, $V' = \{1', 2', 3', 4'\}$ and $E \subseteq V \times V'$, $E = \{(1, 1'), (2, 2'), (3, 4'), (4, 3')\}$. On the right we have the corresponding adjacency matrix A_G.

Let S_n be the set of all the permutations of n elements. More precisely, S_n is the set of bijections $\sigma : [n] \longrightarrow [n]$. Clearly, $|S_n| = n!$, and it is a well known result from algebra that any permutation $\sigma \in S_n$ can be written as a product of transpositions, that is, permutations that simply exchange two elements in $[n]$, and leave every other element fixed. Any permutation in S_n may be written as a product of transpositions, and although there are many ways to do this (i.e., a representation by transpositions is not unique), the parity of the number of transpositions is constant for any given permutation σ. Let $\text{sgn}(\sigma)$ be 1 or -1, depending on whether the parity of σ is even or odd, respectively.

Recall the Lagrange formula for the determinant:

$$\det(A) = \sum_{\sigma \in S_n} \text{sgn}(\sigma) \prod_{i=1}^{n} A_{i\sigma(i)}. \tag{6.1}$$

Problem 6.1. *Let $G = (V \cup V', E)$ be a graph where $n = |V| = |V'|$ and $E \subseteq V \times V'$. Then, the graph G has a* perfect matching *(i.e., each vertex in V can be paired with a unique vertex in V') iff it is the case that $\det(A_G) = \sum_{\sigma \in S_n} \text{sgn}(\sigma) \prod_{i=1}^{n} (A_G)_{i\sigma(i)} \neq 0$.*

Since $|S_n| = n!$, computing the summation over all the σ in S_n, as in (6.1), is computationally very expensive, so we randomly assign values to the x_{ij}'s. The integer determinant, unlike the symbolic determinant, can be computed very efficiently—for example with Berkowitz's algorithm. Let $A_G(x_1, \ldots, x_m)$, $m = |E_G|$, be A_G with its variables renamed to $x_1, \ldots, x_m$. Note that $m \leq n^2$ and each x_l represents some x_{ij}. We obtain a randomized algorithm for the perfect matching problem—see algorithm 6.1.

Algorithm 6.1 Perfect matching

Choose m random integers $i_1, \ldots, i_m$ in $\{1, \ldots, M\}$ where $M = 2m$

compute the integer determinant of $A_G(i_1, \ldots, i_m)$

if $\det(A_G(i_1, \ldots, i_m)) \neq 0$ **then**

      return yes, G has a perfect matching

else

      return no, G *probably* has no perfect matching

end if

Algorithm 6.1 is a polynomial time *Monte Carlo algorithm*: "yes" answers are reliable and final, while "no" answers are in danger of a *false negative*. The false negative can arise as follows: G may have a perfect matching, but $(i_1, \ldots, i_m)$ may happen to be a root of the polynomial $\det(A_G(x_1, \ldots, x_m))$. However, the probability of a false negative (i.e., the probability of error) can be made negligibly small, as we shall see shortly.

In line 1 of algorithm 6.1 we say, somewhat enigmatically, "choose m random numbers." How do we "choose" these random numbers? It turns out that the answer to this question is not easy, and obtaining a source of randomness is the Achilles heel of randomized algorithms. We have the science of *pseudo-random number generators* at our disposal, and other approaches, but this formidable topic lies outside the scope of this book, and so we shall naïvely assume that we have a "some source of randomness."

We want to show the correctness of our randomized algorithm, so we need to show that the probability of error is negligible. We start with the Schwarz-Zipper lemma.

Lemma 6.2 (Schwarz-Zippel). *Consider polynomials over $\mathbb{Z}$, and let $p(x_1, \ldots, x_m) \neq 0$ be a polynomial, where the degree of each variable is $\leq d$ (when the polynomial is written out as a sum of monomials), and let $M > 0$. Then the number of m-tuples $(i_1, \ldots, i_m) \in \{1, 2, \ldots, M\}^m$ such that $p(i_1, \ldots, i_m) = 0$ is $\leq mdM^{m-1}$.*

Proof.   Induction on m (the number of variables). If $m = 1$, $p(x_1)$ can have at most $d = 1 \cdot d \cdot M^{1-1}$ many roots, by the Fundamental Theorem of Algebra.

Suppose the lemma holds for $(m - 1)$, and now we want to give an upper bound of mdM^{m-1} on the number of tuples $(i_1, \ldots, i_m)$ such that $p(i_1, \ldots, i_m) = 0$. First we write $p(x_1, \ldots, x_m)$ as $y_d x_m^d + \cdots + y_0 x_m^0$, where each coefficient $y_i = y_i(x_1, \ldots, x_{m-1}) \in \mathbb{Z}[x_1, \ldots, x_{m-1}]$.

So how many tuples $(i_1, \ldots, i_m)$ such that $p(i_1, \ldots, i_m) = 0$ are there? We partition such tuples into two sets: those that set $y_d = 0$ and those that do not. The result is bounded above by the sum of the upper bounds of the two sets; we now give those upper bounds.

Set 1. By the induction hypothesis, y_d is zero for at most $(m - 1)dM^{m-2}$ many $(i_1, \ldots, i_{m-1})$ tuples, and x_m can take M values, and so $p(x_1, \ldots, x_m)$ is zero for at most $(m - 1)dM^{m-1}$ tuples. Note that we are over-counting here; we are taking *all* tuples that set $y_d = 0$.

Set 2. For each combination of M^{m-1} values for $x_1, \ldots, x_{m-1}$, there are at most d roots of the resulting polynomial (again by the Fundamental Theorem of Algebra), i.e., dM^{m-1}. Note that again we are over-counting as some of those settings to the $x_1, \ldots, x_m$ will result in $y_d = 0$.

Adding the two estimates gives us mdM^{m-1}.   □

Lemma 6.3. *Algorithm 6.1 is correct.*

Proof.   We want to show that algorithm 6.1 for perfect matching is a reliable Monte Carlo algorithm, which means that "yes" answers are 100% correct, while "no" answers admit a negligible probability of error.

If the algorithm answers "yes," then $\det(A_G(i_1, \ldots, i_m)) \neq 0$ for some randomly selected $i_1, \ldots, i_m$, but then the symbolic determinant $\det(A_G(x_1, \ldots, x_m)) \neq 0$, and so, by problem 6.1, G has a perfect matching. So "yes" answers indicate with absolute certainty that there is a perfect matching.

Suppose that the answer is "no." Then we apply lemma 6.2 to $\det(A_G(x_1, \ldots, x_m))$, with $M = 2m$, and obtain that the probability of a false negative is

$$\leq \frac{m \cdot d \cdot M^{m-1}}{M^m} = \frac{m \cdot 1 \cdot (2m)^{m-1}}{(2m)^m} = \frac{m}{2m} = \frac{1}{2}.$$

Now suppose we perform "many independent experiments," meaning that we run algorithm 6.1 k many times, each times choosing a random set $i_1, \ldots, i_m$. Then, if the answer *always* comes zero we know that the probability of error is $\leq \left(\frac{1}{2}\right)^k = \frac{1}{2^k}$. For $k = 100$, the error becomes *negligible*. □

In the last paragraph of the proof of lemma 6.3 we say that we run algorithm 6.1 k many times, and so bring down the probability of error to being less than $\frac{1}{2^k}$, which for $k = 100$ is truly negligible. Running the algorithm k times to get the answer is called *amplification* (because we decrease drastically the probability of error, and so amplify the certainty of having a correct answer); note that the beauty of this approach is that while we run the algorithm only k times, the probability of error goes down exponentially quickly to $\frac{1}{2^k}$. Just to put things in perspective, if $k = 100$, then $\frac{1}{2^{100}}$ is so minuscule that by comparison the probability of earth being hit by a large meteor—while running the algorithm—is a virtual certainty (and being hit by a large meteor would spear anyone the necessity to run algorithms in the first place).

Problem 6.4. *Show how to use algorithm 6.1 to actually find a perfect matching.*

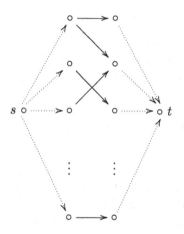

Fig. 6.2   Reduction of perfect matching to max flow.

Perfect matching can be easily reduced[1] to a "max flow problem": as an example, consider the perfect matching problem given in figure 6.1; add two new nodes s, t, and connect s to all the nodes in the left-column of the matching problem, and connect t to all the nodes in the right-column of the matching problem, and give each edge a capacity of 1, and ask if there is a flow $\geq n$ (where n is the number of nodes in each of the two components

---

[1]Recall that we have examined briefly the idea of *reductions* on page 62.

of the given bipartite graph) from s to t; see figure 6.2.

As the max flow problem can be solved in polynomial time without using randomness, it follows that perfect matching can also be solved in polynomial time without randomness. Still, the point of this section was to exhibit a simple randomized algorithm, and that we have accomplished.

6.2   Pattern matching

In this section we design a randomized algorithm for pattern matching. Consider the set of strings over $\{0,1\}$, and let $M : \{0,1\} \longrightarrow M_{2\times 2}(\mathbb{Z})$, that is, M is a map from strings to 2×2 matrices over the integers ($\mathbb{Z}$) defined as follows:

$$M(\varepsilon) = \begin{bmatrix} 1 & 0 \\ 0 & 1 \end{bmatrix}; \quad M(0) = \begin{bmatrix} 1 & 0 \\ 1 & 1 \end{bmatrix}; \quad M(1) = \begin{bmatrix} 1 & 1 \\ 0 & 1 \end{bmatrix},$$

and for strings $x, y \in \{0,1\}^*$, $M(xy) = M(x)M(y)$, where the operation on the left-hand side is concatenation of strings, and the operation on the right-hand side is multiplication of matrices.

Problem 6.5. *Show that $M(x)$ is well defined, that is, no matter how we evaluate M on x we always get the same result. Also show that M is one-to-one.*

Problem 6.6. *Show that for $x \in \{0,1\}^n$, the entries of $M(x)$ are bounded by the n-th Fibonacci number. Recall that Fibonacci numbers were defined in problem 1.5.*

By considering the matrices $M(x)$ modulo a suitable prime p, i.e., by taking all the entries of $M(x)$ modulo a prime p, we perform efficient randomized pattern matching. We wish to determine whether x is a substring of y, where $|x| = n$, $|y| = m$, $n \leq m$. Define

$$y(i) = y_i y_{i+1} \cdots y_{n+i-1},$$

for appropriate $i \in \{1, \ldots, m - n + 1\}$. Select a prime $p \in \{1, \ldots, nm^2\}$, and let $A = M(x) \pmod{p}$ and $A(i) = M(y(i)) \pmod{p}$. Note that

$$A(i + 1) = M^{-1}(y_i)A(i)M(y_{n+i}) \pmod{p},$$

which makes the computation of subsequent $A(i)$'s efficient.

What is the probability of getting a false positive? It is the probability that $A(i) = M(y(i)) \pmod{p}$ even though $A(i) \neq M(y(i))$. This is less

---

Algorithm 6.2 Pattern matching

Pre-condition: $x, y \in \{0,1\}^*$, $|x| = n$, $|y| = m$ and $n \le m$

1: select a random prime $p \le nm^2$
2: $A \longleftarrow M(x) \pmod{p}$
3: $B \longleftarrow M(y(1)) \pmod{p}$
4: **for** $i = 1, \ldots, m - n + 1$ **do**
5:     **if** $A = B$ **then**
6:         **if** $x = y(i)$ **then**
7:             **return** found a match at position i
8:         **end if**
9:     **end if**
10:     $B \longleftarrow M^{-1}(y_i) \cdot B \cdot M(y_{n+i})$
11: **end for**

---

than the probability that $p \in \{1, \ldots, nm^2\}$ divides a (non-zero) entry in $A(i) - M(y(i))$. Since these entries are bounded by $F_n < 2^n$, less than n distinct primes can divide any of them. On the other hand, there are $\pi(nm^2) \approx (nm^2)/(\log(nm^2))$ primes in $\{1, \ldots, nm^2\}$ (by the Prime Number Theorem). So the probability of a false positive is $O(1/m)$.

Note that algorithm 6.2 has no error; it is randomized, but all potential answers are checked for a *false positive* (in line 6). Checking for these potential candidates is called *fingerprinting*. The idea of fingerprinting is to check only those substrings that "look" like good candidates, making sure that when we "sniff" for a candidate we never miss the solution (in this case, if $x = y(i)$, for some i, then $y(i)$ will always be a candidate). On the other hand, there may be j's such that $x \ne y(j)$ and yet they are candidates; but the probability of that is small. The use of randomness in algorithm 6.2 just lowers the average time complexity of the procedure; such algorithms are called *Las Vegas algorithms*.

6.3 Primality testing

One way to determine whether a number p is prime, is to try all possible numbers $n < p$, and check if any are divisors[2]. Obviously, this brute force procedure has exponential time complexity in the length of p, and so it has a prohibitive time cost. Although a polytime (deterministic) algorithm

---

[2]This section requires a little bit of number theory; see appendix A for all the necessary background.

for primality is now known (see [Agrawal *et al.* (2004)]), the Rabin-Miller randomized algorithm for primality testing is simpler and more efficient, and therefore still used in practice.

Fermat's Little theorem (see theorem A.5 in appendix A) provides a "test" of sorts for primality, called the Fermat test; the Rabin-Miller algorithm (algorithm 6.3) is based on this test. When we say that p passes the Fermat test at a, what we mean is that $a^{(p-1)} \equiv 1 \pmod{p}$. Thus, all primes pass the Fermat test for all $a \in \mathbb{Z}_p - \{0\}$.

Unfortunately, there are also composite numbers n that pass the Fermat tests for every $a \in \mathbb{Z}_n^*$; these are the so called *Carmichael numbers*, for example, 561, 1105, 1729, etc.

Lemma 6.7. *If p is a composite non-Carmichael number, then it passes at most half of the tests in $\mathbb{Z}_p^*$. That is, if p is a composite non-Carmichael number, then for at most half of the a's in the set $\mathbb{Z}_p^*$ it is the case that $a^{(p-1)} \equiv 1 \pmod{p}$.*

Proof.   Call a a *witness* (for p) if it fails the Fermat test for p, that is, a is a witness if $a^{(p-1)} \not\equiv 1 \pmod{p}$. Let $S \subseteq \mathbb{Z}_p^*$ consist of those elements $a \in \mathbb{Z}_p^*$ for which $a^{p-1} \equiv 1 \pmod{p}$. It is easy to check that S is in fact a subgroup of $\mathbb{Z}_p^*$. Therefore, by Lagrange theorem (theorem A.8), $|S|$ must divide $|\mathbb{Z}_p^*|$. Suppose now that there exists an element $a \in \mathbb{Z}_p^*$ for which $a^{p-1} \not\equiv 1 \pmod{p}$. Then, $S \neq \mathbb{Z}_p^*$, so the next best thing it can be is "half" of $\mathbb{Z}_p^*$, so $|S|$ must be at most half of $|\mathbb{Z}_p^*|$. $\square$

Problem 6.8. *Give an alternative proof of lemma 6.7 without using group theory.*

A number is *pseudoprime* if it is either prime or Carmichael. The last lemma suggests an algorithm for pseudoprimes: on input p, check whether $a^{(p-1)} \equiv 1 \pmod{p}$ for some random $a \in \mathbb{Z}_p - \{0\}$. If p fails this test (i.e., $a^{(p-1)} \not\equiv 1 \pmod{p}$), then p is composite for sure. If p passes the test, then p is probably pseudoprime. We show that the probability of error in this case is $\leq \frac{1}{2}$. Suppose p is not pseudoprime. If $\gcd(a,p) \neq 1$, then $a^{(p-1)} \not\equiv 1 \pmod{p}$, so assuming that p passed the test, it must be the case that $\gcd(a,p) = 1$, and so $a \in \mathbb{Z}_p^*$. But then, by lemma 6.7, at least half of the elements of $\mathbb{Z}_p^*$ are witnesses of non-pseudoprimeness.

Problem 6.9. *Show that if $\gcd(a,p) \neq 1$ then $a^{(p-1)} \not\equiv 1 \pmod{p}$.*

The informal algorithm for pseudoprimeness described in the paragraph above is the basis for the Rabin-Miller algorithm which we discuss next.

The Rabin-Miller algorithm extends the pseudoprimeness test to deal with Carmichael numbers.

Algorithm 6.3 Rabin-Miller

1: If $n = 2$, accept; if n is even and $n > 2$, reject.
2: Choose at random a positive a in $\mathbb{Z}_n$.
3: **if** $a^{(n-1)} \not\equiv 1 \pmod n$ **then**
4:      reject
5: **else**
6:      Find s, h such that s is odd and $n - 1 = s2^h$
7:      Compute the sequence $a^{s \cdot 2^0}, a^{s \cdot 2^1}, a^{s \cdot 2^2}, \ldots, a^{s \cdot 2^h} \pmod n$
8:      **if** all elements in the sequence are 1 **then**
9:           accept
10:      **else if** the last element different from 1 is -1 **then**
11:           accept
12:      **else**
13:           reject
14:      **end if**
15: **end if**

Note that this is a polytime (randomized) algorithm: computing powers (mod n) can be done efficiently with repeated squaring—for example, if $(n-1)_b = c_r \ldots c_1 c_0$, then compute

$$a_0 = a, a_1 = a_0^2, a_2 = a_1^2, \ldots, a_r = a_{r-1}^2 \pmod n,$$

and so $a^{n-1} = a_0^{c_0} a_1^{c_1} \cdots a_r^{c_r} \pmod n$. Thus obtaining the powers in lines 6 and 7 is not a problem.

Theorem 6.10. *If n is a prime then the Rabin-Miller algorithm accepts it; if n is composite, then the algorithm rejects it with probability $\geq \frac{1}{2}$.*

Proof. If n is prime, then by Fermat's Little theorem $a^{(n-1)} \equiv 1 \pmod n$, so line 4 cannot reject n. Suppose that line 13 rejects n; then there exists a boy b in $\mathbb{Z}_n$ such that $b \not\equiv \pm 1 \pmod n$ and $b^2 \equiv 1 \pmod n$. Therefore, $b^2 - 1 \equiv 0 \pmod n$, and hence

$$(b-1)(b+1) \equiv 0 \pmod n.$$

Since $b \not\equiv \pm 1 \pmod n$, both $(b-1)$ and $(b+1)$ are strictly between 0 and n, and so a prime n cannot divide their product. This gives a contradiction, and therefore no such b exists, and so line 13 cannot reject n.

If n is an odd composite number, then we say that a is a *witness* (of compositness) for n if the algorithm rejects on a. We show that if n is an odd composite number, then at least half of the a's in $\mathbb{Z}_n$ are witnesses. The distribution of those witnesses in $\mathbb{Z}_n$ appears to be very irregular, but if we choose our a at random, we hit a witness with probability $\geq \frac{1}{2}$.

Because n is composite, either n is the power of an odd prime, or n is the product of two odd co-prime numbers. This yields two cases.

Case 1. Suppose that $n = q^e$ where q is an odd prime and $e > 1$. Set $t := 1 + q^{e-1}$. From the binomial expansion of t^n we obtain:

$$t^n = (1 + q^{e-1})^n = 1 + nq^{e-1} + \sum_{l=2}^{n} \binom{n}{l}(q^{e-1})^l, \qquad (6.2)$$

and therefore $t^n \equiv 1 \pmod{n}$. If $t^{n-1} \equiv 1 \pmod{n}$, then $t^n \equiv t \pmod{n}$, which from the observation about t and t^n is not possible, hence t is a line 4 witness. But the set of line 4 non-witnesses, $S_1 := \{a \in \mathbb{Z}_n | a^{(n-1)} \equiv 1 \pmod{n}\}$, is a subgroup of $\mathbb{Z}_n^*$, and since it is not equal to $\mathbb{Z}_n^*$ (t is not in it), by Lagrange's theorem S_1 is at most half of $\mathbb{Z}_n^*$, and so it is at most half of $\mathbb{Z}_n$.

Case 2. Suppose that $n = qr$, where q, r are co-prime. Among all line 13 non-witnesses, find a non-witness for which the -1 appears in the largest position in the sequence in line 7 of the algorithm (note that -1 is a line 13 non-witness, so the set of these non-witnesses is not empty). Let x be such a non-witness and let j be the position of -1 in its sequence, where the positions are numbered starting at 0; $x^{s \cdot 2^j} \equiv -1 \pmod{n}$ and $x^{s \cdot 2^{j+1}} \equiv 1 \pmod{n}$. The line 13 non-witnesses are a subset of $S_2 := \{a \in \mathbb{Z}_n^* | a^{s \cdot 2^j} \equiv \pm 1 \pmod{n}\}$, and S_2 is a subgroup of $\mathbb{Z}_n^*$.

By the CRT there exists $t \in \mathbb{Z}_n$ such that

$$\begin{array}{ll} t \equiv x \pmod{q} & \\ & \Rightarrow \quad \begin{array}{l} t^{s \cdot 2^j} \equiv -1 \pmod{q} \\ t^{s \cdot 2^j} \equiv 1 \pmod{r} \end{array} \\ t \equiv 1 \pmod{r} & \end{array}$$

Hence t is a witness because $t^{s \cdot 2^j} \not\equiv \pm 1 \pmod{n}$ but on the other hand $t^{s \cdot 2^{j+1}} \equiv 1 \pmod{n}$.

Problem 6.11. *Show that $t^{s \cdot 2^j} \not\equiv \pm 1 \pmod{n}$.*

Therefore, just as in case 1, we have constructed a $t \in \mathbb{Z}_n^*$ which is not in S_2, and so S_2 can be at most half of $\mathbb{Z}_n^*$, and so at least half of the elements in $\mathbb{Z}_n$ are witnesses. $\qquad\square$

Problem 6.12. *First show that the sets S_1 and S_2 (in the proof of theorem 6.10) are indeed subgroups of $\mathbb{Z}_n^*$, and that in case 2 all non-witnesses* *are contained in* S_2. *Then show that at least half of the elements of* $\mathbb{Z}_n$ *are witnesses when* n *is composite, without using group theory.*

Note that by running the algorithm k times on independently chosen a, we can make sure that it rejects a composite with probability $\geq 1 - \frac{1}{2^k}$ (it will always accept a prime with probability 1). Thus, for $k = 100$ the probability of error, i.e., of a false positive, is negligible.

6.4 Public key cryptography

Cryptography has well known applications to security; for example, we can use our credit cards when purchasing online because, when we send our credit card numbers, they are encrypted, and even though they travel through a public channel, no one, but the intended recipient, can read them. Cryptography has also a fascinating history: from the first uses recorded by Herodotus during the Persian wars five centuries BC, to the exploits at Bletchley Park during WWII—the reader interested in the history of cryptography should read the fascinating book [Singh (1999)].

A *Public Key Cryptosystem (PKC)* consists of three sets: K, the set of (pairs of) *keys*, M, the set of *plaintext* messages, and C, the set of *ciphertext* messages. A pair of keys in K is $k = (k_{\text{priv}}, k_{\text{pub}})$; the *private* (or *secret*) key and the *public* key, respectively. For each k_{pub} there is a corresponding *encryption* function $e_{k_{\text{pub}}} : M \longrightarrow C$ and for each k_{priv} there is a corresponding *decryption* function $d_{k_{\text{priv}}} : C \longrightarrow M$.

The property that the encryption and decryption functions must satisfy is that if $k = (k_{\text{priv}}, k_{\text{pub}}) \in K$, then $d_{k_{\text{priv}}}(e_{k_{\text{pub}}}(m)) = m$ for all $m \in M$. The necessary assumption is that it must be difficult to compute $d_{k_{\text{priv}}}(c)$ just from knowing k_{pub} and c. But, with the additional *trapdoor* information k_{priv} it becomes easy to compute $d_{k_{\text{priv}}}(c)$.

In the following sections we present three different encryption scheme; Diffie-Hellman, which is not really a PKC but rather a way of agreeing on a secret key over an insecure channel, as well as ElGamal and RSA. All three require large primes (in practice about 1,000 bit long); a single prime for Diffie-Hellman and ElGamal, and a pair of primes for RSA. How to go about it? The answer will of course involve the Rabin-Miller algorithm from the previous section.

Here is how we go about it: we know by the Prime Number Theorem that there are about $\pi(n) = n / \log n$ many primes $\leq n$. This means that there are $2^n/n$ primes among n-bit integers, roughly 1 in n, and these primes are fairly uniformly distributed. So we pick an integer at random, in a given range, and apply the Rabin-Miller algorithm to it.

6.4.1 *Diffie-Hellman key exchange*

If p is prime, then one can show—though the proof is difficult and we omit it here—that there exists a $g \in \mathbb{Z}_p^*$ such that $\langle g \rangle = \{g^1, g^2, \ldots, g^{p-1}\} = \mathbb{Z}_p^*$. This g is called a *primitive root* for $\mathbb{Z}_p^*$. Given an $h \in \mathbb{Z}_p^*$, the *Discrete Log Problem (DLP)* is the problem of finding an $x \in \{1, \ldots, p-1\}$ such that $g^x \equiv h \pmod{p}$. That is, $x = \log_g(h)$.

For example, $p = 56609$ is a prime number and $g = 2$ is a generator for $\mathbb{Z}_{56609}^*$, that is $\mathbb{Z}_{56609}^* = \{2^1, 2^2, 2^3, \ldots, 2^{56608}\}$, and $\log_2(38679) = 11235$.

The DLP is assumed to be a difficult problem. We are going to use it to set up a way for Alice and Bob to agree on a secret key over an insecure channel. First Alice and Bob agree on a large prime p and an integer $g \in \mathbb{Z}_p^*$. The numbers p, g are public knowledge, that is, $k_{\text{pub}} = \langle p, g \rangle$.

Then Alice picks a secret a and Bob picks a secret b. Note that Diffie-Hellman is not really a fully-fledged PKC; it is just a way for two parties to agree on a secret value over an insecure channel.

Alice computes $A := g^a \pmod{p}$ and Bob computes $B := g^b \pmod{p}$. Then Alice and Bob exchange A and B over an (insecure) link. On her end, Alice computes $A' := B^a \pmod{p}$ and Bob, on his end, computes $B' := A^b \pmod{p}$. Clearly,

$$A' \equiv_p B^a \equiv_p (g^b)^a \equiv_p g^{ab} \equiv_p (g^a)^b \equiv_p A^b \equiv_p B'.$$

This common value $A' = B'$ is their secret key.

Suppose that Eve is eavesdropping on this exchange. She is capable of gleaning the following information from it: $\langle p, g, g^a \pmod{p}, g^b \pmod{p} \rangle$. Computing $g^{ab} \pmod{p}$ (i.e., $A' = B'$) from this information is known as the *Diffie-Hellman Problem* (DHP), and it is assumed to be difficult when p is a large prime number.

But suppose that Eve has an efficient way of solving the DLP. Then, from $g^a \pmod{p}$ she computes a, and from $g^b \pmod{p}$ she computes b, and now she can easily compute $g^{ab} \pmod{p}$. On the other hand, it is not known if solving DHP efficiently yields an efficient solution for the DLP.

Problem 6.13. *Consider Shank's algorithm—algorithm 6.4. Show that Shank's algorithm computes x, such that $g^x \equiv h \pmod{p}$, in time $O(n \log n)$ that is, in time $O(\sqrt{p} \log(\sqrt{p}))$.*

Algorithm 6.4 Shank's babystep-giantstep

1: On input g, p, h
2: $n = 1 + \lfloor \sqrt{p} \rfloor$
3: Compute list $g^0, g^1, g^2, \ldots, g^n$ (mod p)
4: Compute list $hg^0, hg^{-n}, hg^{-2n}, \ldots, hg^{-n^2}$ (mod p)
5: Find a match between the two lists, say $g^i = hg^{-jn}$
6: **return** $x = jn + i$

6.4.2 *ElGamal*

This is a true PKC. Alice and Bob agree on public p, g. Alice also has a private a and a public $A := g^a$ (mod p). Bob wants to send a message m to Alice, so he creates an *ephemeral* key b, and sends the following pair c_1, c_2 to Alice:

$$\langle c_1 := g^b \quad (\text{mod } p), \qquad c_2 := mA^b \quad (\text{mod } p) \rangle.$$

Then, in order to read the message, Alice computes:

$$c_1^{-a} c_2 \equiv_p g^{-ab} m g^{ab} \equiv_p m.$$

Note that to compute c_1^{-a} Alice first computes the inverse of c_1 (in $\mathbb{Z}_p^*$), which she can do efficiently using the extended Euclid's algorithm (see problem 1.18), and then computes the a-th power of the result.

Problem 6.14. *We say that we can* break *ElGamal, if we have an efficient way for computing m from $\langle p, g, A, c_1, c_2 \rangle$. Show that we can break ElGamal if and only if we can solve the DHP efficiently.*

6.4.3 *RSA*

Choose two odd primes p, q, and set $n = pq$. Choose $k \in \mathbb{Z}_{\phi(n)}^*$, $k > 1$. Advertise f, where $f(m) \equiv m^k$ (mod n). Compute $l = k^{-1}$ (inverse of k in $\mathbb{Z}_{\phi(n)}^*$). Now $\langle n, k \rangle$ are public, and the key l is secret, and so is the function g, where $g(C) \equiv C^l$ (mod n). Note that $g(f(m)) \equiv_n m^{kl} \equiv_n m$.

Problem 6.15. *Show that $m^{kl} \equiv m$ (mod n). In fact there is an implicit assumption about m in order for this to hold; what is this assumption? You may want to read the next paragraph before attempting this problem.*

Note that computing the inverse of k in $\mathbb{Z}_{\phi(n)}^*$, that is l, can be done in polytime using the extended Euclid's algorithm. Just observe that if

$k \in \mathbb{Z}^*_{\phi(n)}$, then $\gcd(k, \phi(n)) = 1$, so $\exists s, t$ such that $sk + t\phi(n) = 1$, and further s, t can be chosen so that s is in $\mathbb{Z}^*_{\phi(n)}$ (first obtain any s, t from the extended Euclid's algorithm, and then just add to s the appropriate number of (positive or negative) multiples of $\phi(n)$ to place it in the set $\mathbb{Z}^*_{\phi(n)}$, and adjust t by the same number of multiples (of opposite sign)); set $l := s$.

Problem 6.16. *Observe that we could break RSA if factoring were easy.*

We now discuss very briefly two issues related to the security of RSA. The first one is that the primes p, q cannot be chosen "close" to each other. Note that

$$n = \left(\frac{p+q}{2}\right)^2 - \left(\frac{p-q}{2}\right)^2.$$

Since p, q are close, we know that $s := \dfrac{p-q}{2}$ is small, and $t := \dfrac{p+q}{2}$ is only slightly larger than $n^{\frac{1}{2}}$, and $t^2 - n = s^2$ is a perfect square. So we try the following candidate values for t:

$$\lceil n^{\frac{1}{2}} \rceil, \lceil n^{\frac{1}{2}} \rceil + 1, \lceil n^{\frac{1}{2}} \rceil + 2, \ldots$$

until $t^2 - n$ is a perfect square s^2. Clearly, if s is small, we will quickly find such a t, and then $p = t + s$ and $q = t - s$.

The second issue is the following: suppose that Eve can compute $\phi(n)$ from n. Then she can easily compute the primes p, q (of course, if she can compute $\phi(n)$ she can directly compute l, and she does not need p, q). To see this note that $\phi(n) = \phi(pq) = (p-1)(q-1)$. Then,

$$p + q = n - \phi(n) + 1$$
$$pq = n,$$
(6.3)

and from these two equations,

$$(x - p)(x - q) = x^2 - (p+q)x + pq = x^2 - (n - \phi(n) + 1)x + n.$$

Thus, we can compute p, q by computing the roots of this last polynomial, and using the quadratic formula $x = (-b \pm \sqrt{b^2 - 4ac})/2a$, we obtain that p, q are

$$\frac{(n - \phi(n) + 1) \pm \sqrt{(n - \phi(n) + 1)^2 - 4n}}{2}.$$

Suppose that Eve is able to compute l from n and k. If Eve knows l, then she knows that whatever $\phi(n)$ is, it divides $kl - 1$, so she has equations (6.3) but with $\phi(n)$ in the first equation replaced by $(kl-1)/a$, for some unknown a. There is a randomized polytime procedure to find the appropriate a, and obtain p, q, but we do not describe it here.

If Eve is able to factor she can obviously break RSA; on the other hand, if Eve can break RSA (by computing l from n, k), then she would be able to factor in randomized polytime. Conceivably, Eve may be able to break RSA *without* computing l, so these observation do not imply that breaking RSA is as hard as factoring.

6.5 Answers to selected problems

Problem 6.4. We use algorithm 6.1 to find perfect matching (if one exists) as follows: pick $1 \in V$, and consider each $(1, i') \in E$ in turn, remove it from G to obtain $G_{1,i'} = ((V - \{1\}) \cup (V' - \{i'\}), E_{1,i'})$, where $E_{1,i'}$ consists of all the edges of E except those adjacent on 1 or i', until for some $i' \in V'$ we obtain a $G_{1,i'}$ for which the algorithm answers "yes." Then we know that there is a perfect matching that matches 1 and i'. Continue with $G_{1,i'}$.

Problem 6.5. $M(x)$ is well defined because matrix multiplication is associative. We now show that $M(x) = M(y)$ implies that $x = y$ (i.e., the map M is one-to-one). Given $M = M(x)$ we can "decode" x uniquely as follows: if the first column of M is greater than the second (where the comparison is made component-wise), then the last bit of x is zero, and otherwise it is 1. Let M' be M where we subtract the smaller column from the larger, and repeat.

Problem 6.6. For a given string x, $M(x_1 x_2 \ldots x_n)$ is such that the "smaller" column is bounded by f_{n-1} and the "larger" column is bounded by f_n. We can show this inductively: the basis case, $x = x_1$, is obvious. For the inductive step, assume it holds for $x \in \{0, 1\}^n$, and show it still holds for $x \in \{0, 1\}^{n+1}$: this is clear as whether x_{n+1} is 0 or 1, one column is added to the other, and the other column remains unchanged.

Problem 6.9. Suppose that $\gcd(a, p) \neq 1$. By proposition A.4 we know that if $\gcd(a, p) \neq 1$, then a does not have a (multiplicative) inverse in $\mathbb{Z}_p$. Thus, it is not possible for $a^{(p-1)} \equiv 1 \pmod{p}$ to be true, since then it would follow that $a \cdot a^{(p-2)} \equiv 1 \pmod{p}$, and hence a would have a (multiplicative) inverse.

Problem 6.11. To see why $t^{s \cdot 2^j} \not\equiv \pm 1 \pmod{n}$ observe the following: suppose that $a \equiv -1 \pmod{q}$ and $a \equiv 1 \pmod{r}$, where $\gcd(q, r) = 1$. Suppose that $n = qr | (a+1)$, then $q | (a+1)$ and $r | (a+1)$, and since $r | (a-1)$

as well, it follows that $r|[(a+1)-(a-1)]$, so $r|2$, so $r=2$, so n must be even, which is not possible since we deal with even n's in line 1 of the algorithm.

Problem 6.12. Showing that S_1, S_2 are subgroups of $\mathbb{Z}_n^*$ is easy; it is obvious in both cases that 1 is there, and closure and existence of inverse can be readily checked.

To give the same proof without group theory, we follow the cases in the proof of theorem 6.10. Let t be the witness constructed in case 1. If d is a (stage 3) non-witness, we have $d^{p-1} \equiv 1 \pmod{p}$, but then $dt \pmod{p}$ is a witness. Moreover, if d_1, d_2 are distinct (stage 3) non-witnesses, then $d_1t \not\equiv d_2t \pmod{p}$. Otherwise, $d_1 \equiv_p d_1 \cdot t \cdot t^{p-1} \equiv_p d_2 \cdot t \cdot t^{p-1} \equiv_p d_2$. Thus the number of (stage 3) witnesses must be at least as large as the number of non-witnesses.

We do the same for case 2; let d be a non-witness. First, $d^{s \cdot 2^j} \equiv \pm 1$ $\pmod{p}$ and $d^{s \cdot 2^{j+1}} \equiv 1 \pmod{p}$ owing to the way that j was chosen. Therefore $dt \pmod{p}$ is a witness because $(dt)^{s \cdot 2^j} \not\equiv \pm 1 \pmod{p}$ and $(dt)^{s \cdot 2^{j+1}} \equiv 1 \pmod{p}$.

Second, if d_1 and d_2 are distinct non-witnesses, $d_1t \not\equiv d_2t \pmod{p}$. The reason is that $t^{s \cdot 2^{j+1}} \equiv 1 \pmod{p}$. Hence $t \cdot t^{s \cdot 2^{j+1}-1} \equiv 1 \pmod{p}$. Therefore, if $d_1t \equiv d_2t \pmod{p}$, then $d_1 \equiv_p d_1t \cdot t^{s \cdot 2^{j+1}-1} \equiv_p d_2 \cdot t^{s \cdot 2^{j+1}-1} \equiv_p d_2$. Thus in case 2, as well, the number of witnesses must be at least as large as the number of non-witnesses.

Problem 6.14. The DHP on input $\langle p, g, g^a \pmod{p}, g^b \pmod{p} \rangle$ outputs $g^{ab} \pmod{p}$, and the ElGamal problem, call it ELGP, on input

$$\langle p, g, A = g^a \pmod{p}, c_1 = g^b \pmod{p}, c_2 = mA^b \pmod{p} \rangle$$

outputs m. We want to show that we can break Diffie-Hellman (i.e., solve DHP efficiently) if and only if we can break ElGamal (i.e., solve ELGP efficiently). The key-word here is efficiently, i.e., in polynomial time.

($\Rightarrow$) Suppose we can solve DHP efficiently. We now give an efficient procedure for solving ELGP: given the input to ELGP, from $A = g^a \pmod{p}$ and $c_1 = g^b \pmod{p}$ we obtain $g^{ab} \pmod{p}$, using the efficient solver for DHP (which we have by assumption). We then use the extended Euclidean algorithm (see problem 1.18) to obtain $(g^{ab})^{-1} \pmod{p}$ (note that the extended Euclidean algorithm is polynomial time), and now we compute:

$$c_2 \cdot (g^{ab})^{-1} \pmod{p} = mg^{ab}(g^{ab})^{-1} \pmod{p} = m \pmod{p} = m$$

where the last equality follows from the fact that we always assume that m is in $\mathbb{Z}_p$.

($\Leftarrow$) Suppose we have an efficient solver for the ELGP. To solve the DHP, we construct the following input to ELGP:

$$\langle p, g, A = g^a \pmod{p}, c_1 = g^b \pmod{p}, c_2 = 1 \rangle.$$

Note that $c_2 = 1 = \underbrace{(g^{ab})^{-1} A^b}_{=m} \pmod{p}$, so using the efficient solver for ELGP we obtain $m = (g^{ab})^{-1} \pmod{p}$, and now using the extended Euclidean algorithm we obtain the inverse of $(g^{ab})^{-1} \pmod{p}$, which is just $g^{ab} \pmod{p}$, so we output that.

Problem 6.15. Why $m^{kl} \equiv_n m$? Observe that $kl = 1 + (-t)\phi(n)$, where $(-t) > 0$, and so $m^{kl} \equiv_n m^{1+(-t)\phi(n)} \equiv_n m \cdot (m^{\phi(n)})^{(-t)} \equiv_n m$, because $m^{\phi(n)} \equiv_n 1$. Note that this last statement does not follow directly from Euler's theorem (theorem A.10), because $m \in \mathbb{Z}_n$, and not necessarily in $\mathbb{Z}_n^*$. Note that to make sure that $m \in \mathbb{Z}_n^*$ it is enough to insist that we have $0 < m < \min\{p, q\}$; so we break a large message into small pieces.

It is interesting to note that we can bypass Euler's theorem, and just use Fermat's Little theorem: we know that $m^{(p-1)} \equiv_p 1$ and $m^{(q-1)} \equiv_q 1$, so $m^{(p-1)(q-1)} \equiv_p 1$ and $m^{(q-1)(p-1)} \equiv_q 1$, thus $m^{\phi(n)} \equiv_p 1$ and $m^{\phi(n)} \equiv_q 1$. This means that $p|(m^{\phi(n)} - 1)$ and $q|(m^{\phi(n)} - 1)$, so, since p, q are distinct primes, it follows that $(pq)|(m^{\phi(n)} - 1)$, and so $m^{\phi(n)} \equiv_n 1$.

Problem 6.16. If factoring integers were easy, RSA would be easily broken: if we were able to factor n, we would obtain the primes p, q, and hence it would be easy to compute $\phi(n) = \phi(pq) = (p-1)(q-1)$, and from this we obtain l, the inverse of k.

6.6 Notes

This chapter is based on chapter 8 in [Soltys (2009)]. For a discussion of Berkowitz's algorithm see [Berkowitz (1984); Soltys (2002)].

Most algorithms books provide a chapter on the *Min-Cut Max-Flow* problem mentioned in the context of perfect matching; see for example chapter 7 in [Kleinberg and Tardos (2006)].

For a discussion on generating random numbers see, for example, chapter 7 in [Press *et al.* (2007)].

Credit for inventing the Monte Carlo method often goes to Stanisław Ulam, a Polish born mathematician who worked for John von Neumann on the United States Manhattan Project during World War II. Ulam is primarily known for designing the hydrogen bomb with Edward Teller in 1951. He invented the Monte Carlo method in 1946 while pondering the probabilities of winning a card game of solitaire.

Section 6.2 is based on [Karp and Rabin (1987)].

It was the randomized test for primality that stirred interest in randomized computation in the late 1970's. Historically, the first randomized algorithm for primality was given by [Solovay and Strassen (1977)]; a good exposition of this algorithm, with all the necessary background, can be found in §11.1 in [Papadimitriou (1994)], and another in §18.5 in [von zur Gathen and Gerhard (1999)].

R. D. Carmichael first noted the existence of the Carmichael numbers in 1910, computed fifteen examples, and conjectured that though they are infrequent there were infinitely many. In 1956, Erdös sketched a technique for constructing large Carmichael numbers ([Hoffman (1998)]), and a proof was given by [Alford *et al.* (1994)] in 1994.

The first three Carmichael numbers are 561, 1105, 1729, where the last number shown on this list is called the Hardy-Ramanujan number, after a famous anecdote of the British mathematician G. H. Hardy regarding a hospital visit to the Indian mathematician Srinivasa Ramanujan. Hardy wrote: *I remember once going to see him when he was ill at Putney. I had ridden in taxi cab number 1729 and remarked that the number seemed to me rather a dull one, and that I hoped it was not an unfavorable omen. "No," he replied, "it is a very interesting number; it is the smallest number expressible as the sum of two cubes in two different ways.".*

Section 6.4 is based on material from [Hoffstein *et al.* (2008)] and [Delfs and Knebl (2007)].

RSA is named after the first letters of the last names of its inventors: Ron **R**ivest, Adi **S**hamir, and Leonard **A**dleman.

Appendix A

Number Theory and Group Theory

In this section we work with the set of integers and natural numbers

$$\mathbb{Z} = \{\ldots, -3, -2, -1, 0, 1, 2, 3, \ldots\}, \quad \mathbb{N} = \{0, 1, 2, \ldots\}.$$

We say that x *divides* y, and write $x|y$ if $y = qx$. If $x|y$ we say that x is *divisor* (also *factor*) of y. Using the terminology from section 1.3, $x|y$ if and only if $y = \operatorname{div}(x, y) \cdot x$. We say that a number p is *prime* if its only divisors are itself and 1.

Claim A.1. *If p is a prime, and $p|a_1 a_2 \ldots a_n$, then $p|a_i$ for some i.*

Proof. It is enough to show that if $p|ab$ then $p|a$ or $p|b$. Let $g = \gcd(a, p)$. Then $g|p$, and since p is a prime, there are two cases. Case 1, $g = p$, then since $g|a$, $p|a$. Case 2, $g = 1$, so there exist u, v such that $au + pv = 1$ (see problem 1.18), so $abu + pbv = b$. Since $p|ab$, and $p|p$, it follows that $p|(abu + pbv)$, so $p|b$. □

Theorem A.2 (Fundamental Theorem of Arithmetic). *For $a \geq 2$, $a = p_1^{e_1} p_2^{e_2} \cdots p_r^{e_r}$, where p_i are prime numbers, and other than rearranging primes, this factorization is unique.*

Proof. We first show the existence of the factorization, and then its uniqueness. The proof of existence is by complete induction; the basis case is $a = 2$, where 2 is a prime. Consider an integer $a > 2$; if a is prime then it is its own factorization (just as in the basis case). Otherwise, if a is composite, then $a = b \cdot c$, where $1 < b, c < a$; apply the induction hypothesis to b and c.

To show uniqueness suppose that $a = p_1 p_2 \ldots p_s = q_1 q_2 \ldots q_t$ where we have written out all the primes, that is, instead of writing p^e we write $p \cdot p \cdots p$, e times. Since $p_1|a$, it follows that $p_1|q_1 q_2 \ldots q_t$. So $p_1|q_j$ for some

j, by claim A.1, but then $p_1 = q_j$ since they are both primes. Now delete p_1 from the first list and q_j from the second list, and continue. Obviously we cannot end up with a product of primes equal to 1, so the two list must be identical.   □

Let $m \geq 1$ be an integer. We say that a and b are *congruent modulo* m, and write $a \equiv b \pmod{m}$ (or sometimes $a \equiv_m b$) if $m|(a-b)$. Another way to say this is that a and b have the same remainder when divided by m; using the terminology of section 1.3, we can say that $a \equiv b \pmod{m}$ if and only if $\mathrm{rem}(m, a) = \mathrm{rem}(m, b)$.

Problem A.3. *Show that if $a_1 \equiv_m a_2$ and $b_1 \equiv_m b_2$, then $a_1 \pm b_1 \equiv_m a_2 \pm b_2$ and $a_1 \cdot b_1 \equiv_m a_2 \cdot b_2$.*

Proposition A.4. *If $m \geq 1$, then $a \cdot b \equiv_m 1$ for some b if and only if $\gcd(a, m) = 1$.*

Proof. ($\Rightarrow$) If there exists a b such that $a \cdot b \equiv_m 1$, then we have $m|(ab-1)$ and so there exists a c such that $ab - 1 = cm$, i.e., $ab - cm = 1$. And since $\gcd(a, m)$ divides both a and m, it also divides $ab - cm$, and so $\gcd(a, m)|1$ and so it must be equal to 1.

($\Leftarrow$) Suppose that $\gcd(a, m) = 1$. By the extended Euclid's algorithm (see problem 1.18) there exist u, v such that $au + mv = 1$, so $au - 1 = -mv$, so $m|(au - 1)$, so $au \equiv_m 1$. So let $b = u$.   □

Let $\mathbb{Z}_m = \{0, 1, 2, \ldots, m-1\}$. We call $\mathbb{Z}_m$ the set of integers modulo m. To add or multiply in the set $\mathbb{Z}_m$, we add and multiply the corresponding integers, and then take the reminder of the division by m as the result. Let $\mathbb{Z}_m^* = \{a \in \mathbb{Z}_m | \gcd(a, m) = 1\}$. By proposition A.4 we know that $\mathbb{Z}_m^*$ is the subset of $\mathbb{Z}_m$ consisting of those elements which have multiplicative inverses in $\mathbb{Z}_m$.

The function $\phi(n)$ is called the *Euler totient function*, and it is the number of elements less than n that are co-prime to n, i.e., $\phi(n) = |\mathbb{Z}_n^*|$. If we are able to factor, we are also able to compute $\phi(n)$: suppose that $n = p_1^{k_1} p_2^{k_2} \cdots p_l^{k_l}$, then it is not hard to see that $\phi(n) = \prod_{i=1}^{l} p_i^{k_i-1}(p_i - 1)$.

Theorem A.5 (Fermat's Little Theorem). *Let p be a prime number and $\gcd(a, p) = 1$. Then $a^{p-1} \equiv 1 \pmod{p}$.*

Proof. For any a such that $\gcd(a, p) = 1$ the following numbers

$$1 \cdot a, 2 \cdot a, 3 \cdot a, \ldots, (p-1) \cdot a, \tag{A.1}$$

all taken mod p, are pairwise distinct. To see this suppose that $ja \equiv ka$ (mod p). Then $(j-k)a \equiv 0$ (mod p), and so $p|(j-k)a$. But since by assumption $\gcd(a,p) = 1$, it follows that $p \nmid a$, and so by claim A.1 it must be the case that $p|(j-k)$. But since $j, k \in \{1, 2, \ldots, p-1\}$, it follows that $-(p-2) \leq j - k \leq (p-2)$, so $j - k = 0$, i.e., $j = k$.

Thus the numbers in (A.1) are just a re-ordering of $\{1, 2, \ldots, p-1\}$. Therefore

$$a^{p-1}(p-1)! \equiv_p \prod_{j=1}^{p-1} j \cdot a \equiv_p \prod_{j=1}^{p-1} j \equiv_p (p-1)!. \tag{A.2}$$

Since all the numbers in $\{1, 2, \ldots, p-1\}$ have inverses in $\mathbb{Z}_p$, as $\gcd(i,p) = 1$ for $1 \leq i \leq p-1$, their product also has an inverse. That is, $(p-1)!$ has an inverse, and so multiplying both sides of (A.2) by $((p-1)!)^{-1}$ we obtain the result. $\square$

Problem A.6. *Given a second proof of Fermat's Little theorem using the binomial expansion, i.e., $(x+y)^n = \sum_{j=0}^{n} \binom{n}{j} x^j y^{n-j}$ applied to $(a+1)^p$.*

We say that $(G, *)$ is a *group* if G is a set and $*$ is an operation, such that if $a, b \in G$, then $a * b \in G$ (this property is called *closure*). Furthermore, the operation $*$ has to satisfy the following three properties:

(1) *identity law:* There exists an $e \in G$ such that $e * a = a * e = a$ for all $a \in G$.
(2) *inverse law:* For every $a \in G$ there exists an element $b \in G$ such that $a * b = b * a = e$. This element b is called an *inverse* and it can be shown that it is unique; hence it is often denoted as a^{-1}.
(3) *associative law:* For all $a, b, c \in G$, we have $a * (b * c) = (a * b) * c$.

If $(G, *)$ also satisfies the *commutative law*, that is, if for all $a, b \in G$, $a * b = b * a$, then it is called a *commutative* or *Abelian group*.

Typical examples of groups are $(\mathbb{Z}_p, +)$ (integers mod p under addition) and $(\mathbb{Z}_p^*, \cdot)$ (integers mod p under multiplication). Note that both these groups are Abelian. These are, of course, the two groups of concern for us; but there are many others: $(\mathbb{Q}, +)$ is an infinite group (rationals under addition), $\text{GL}(n, \mathbb{F})$ (which is the group of $n \times n$ invertible matrices over a field $\mathbb{F}$), and S_n (the *symmetric group* over n elements, consisting of permutations of $[n]$ where $*$ is function composition).

We let $|G|$ denote the number of elements in G (note that G may be infinite, but we are concerned mainly with finite groups). If $g \in G$ and

$x \in \mathbb{N}$, then $g^x = g * g * \cdots * g$, x times. Note that if the operation $*$ is clear from the context then we write ab instead of $a * b$.

Suppose that G is a finite group and $a \in G$; then the smallest $d \in \mathbb{N}$ such that $a^d = e$ is called the *order* of a, and it is denoted as $\mathrm{ord}_G(a)$ (or just $\mathrm{ord}(a)$ if the group G is clear from the context).

Proposition A.7. *If G is a finite group, then for all $a \in G$ there exists a $d \in \mathbb{N}$ such that $a^d = e$. If $d = \mathrm{ord}_G(a)$, and $a^k = e$, then $d|k$.*

Proof.   Consider the list $a^1, a^2, a^3, \ldots$. If G is finite there must exist $i < j$ such that $a^i = a^j$. Then, $(a^{-1})^i$ applied to both sides yields $a^{i-j} = e$. Let $d = \mathrm{ord}(a)$ (by the LNP we know that it must exist!). Suppose that $k \geq d$, $a^k = e$. Write $k = dq + r$ where $0 \leq r < d$. Then $e = a^k = a^{dq+r} = (a^d)^q a^r = a^r$. Since $a^d = e$ it follows that $a^r = e$, contradicting the minimality of $d = \mathrm{ord}(a)$, unless $r = 0$.   □

If $(G, *)$ is a group we say that H is a *subgroup* of G, and write $H \leq G$ if $H \subseteq G$ and H is closed under $*$. That is, H is a subset of G, and H is itself a group. Note that for any G it is always the case that $\{e\} \leq G$ and $G \leq G$; these two are called the *trivial subgroups* of G. If $H \leq G$ and $g \in G$, then gH is called a *left coset of G*, and it is simply the set $\{gh | h \in H\}$. Note that gH is not necessarily a subgroup of G.

Theorem A.8 (Lagrange). *If G is a finite group and $H \leq G$, then $|H|$ divides $|G|$, i.e., the order of H divides the order of G.*

Proof.   If $g_1, g_2 \in G$, then the two cosets $g_1 H$ and $g_2 H$ are either identical or $g_1 H \cap g_2 H = \emptyset$. To see this, suppose that $g \in g_1 H \cap g_2 H$, so $g = g_1 h_1 = g_2 h_2$. In particular, $g_1 = g_2 h_2 h_1^{-1}$. Thus, $g_1 H = (g_2 h_2 h_1^{-1})H$, and since it can be easily checked that $(ab)H = a(bH)$ and that $hH = H$ for any $h \in H$, it follows that $g_1 H = g_2 H$.

Therefore, for a finite $G = \{g_1, g_2, \ldots, g_n\}$, the collection of sets $\{g_1 H, g_2 H, \ldots, g_n H\}$ is a partition of G into subsets that are either disjoint or identical; from among all subcollections of identical cosets we pick a representative, so that $G = g_{i_1} H \cup g_{i_2} H \cup \cdots \cup g_{i_m} H$, and so $|G| = m|H|$, and we are done.   □

Problem A.9. *If G is a group, and $\{g_1, g_2, \ldots, g_k\} \subseteq G$, then the set $\langle g_1, g_2, \ldots, g_k \rangle$ is defined as follows*

$$\{x_1 x_2 \cdots x_p | p \in \mathbb{N}, x_i \in \{g_1, g_2, \ldots, g_k, g_1^{-1}, g_2^{-1}, \ldots, g_k^{-1}\}\}.$$

Show that $\langle g_1, g_2, \ldots, g_k \rangle \leq G$, *and it is called the subgroup* generated *by* $\{g_1, g_2, \ldots, g_k\}$. *Also show that when* G *is finite* $|\langle g \rangle| = \mathrm{ord}_G(g)$.

Theorem A.10 (Euler). *For every* n *and every* $a \in \mathbb{Z}_n^*$, *that is, for every pair* a, n *such that* $\gcd(a, n) = 1$, *we have* $a^{\phi(n)} \equiv 1 \pmod{n}$.

Proof. First it is easy to check that $(\mathbb{Z}_n^*, \cdot)$ is a group. Then by definition $\phi(n) = |\mathbb{Z}_n^*|$, and since $\langle a \rangle \leq \mathbb{Z}_n^*$, it follows by Lagrange's theorem that $\mathrm{ord}(a) = |\langle a \rangle|$ divides $\phi(n)$. $\square$

Note that Fermat's Little theorem is an immediate consequence of Euler's Theorem, since when p is a prime, $\mathbb{Z}_p^* = \mathbb{Z}_p - \{0\}$, and $\phi(p) = (p - 1)$.

Theorem A.11 (Chinese Remainder). *Given two sets of numbers of equal size,* $r_0, r_1, \ldots, r_n$, *and* $m_0, m_1, \ldots, m_n$, *such that*

$$0 \leq r_i < m_i \qquad 0 \leq i \leq n \qquad (A.3)$$

and $\gcd(m_i, m_j) = 1$ *for* $i \neq j$, *then there exists an* r *such that* $r \equiv r_i \pmod{m_i}$ *for* $0 \leq i \leq n$.

Proof. The proof we give is by counting; we show that the distinct values of r, $0 \leq r < \Pi m_i$, represent distinct sequences. To see that, note that if $r \equiv r' \pmod{m_i}$ for all i, then $m_i | (r - r')$ for all i, and so $(\Pi m_i) | (r - r')$, since the m_i's are pairwise co-prime. So $r \equiv r' \pmod{(\Pi m_i)}$, and so $r = r'$ since both $r, r' \in \{0, 1, \ldots, (\Pi m_i) - 1\}$.

But the total number of sequences $r_0, \ldots, r_n$ such that (A.3) holds is precisely Πm_i. Hence every such sequence must be a sequence of remainders of some r, $0 \leq r < \Pi m_i$. $\square$

Problem A.12. *The proof of theorem A.11 (CRT) is non-constructive. Show how to obtain efficiently the* r *that meets the requirement of the theorem, i.e., in polytime in* n—*so in particular not using brute force search.*

Given two groups $(G_1, *_1)$ and $(G_2, *_2)$, a mapping $h : G_1 \longrightarrow G_2$ is a *homomorphism* if it respects the operation of the groups; formally, for all $g_1, g_1' \in G_1$, $h(g_1 *_1 g_1') = h(g_1) *_2 h(g_1')$. If the homomorphism h is also a bijection, then it is called an *isomorphism*. If there exists an isomorphism between two groups G_1 and G_2, we call them *isomorphic*, and write $G_1 \cong G_2$.

If $(G_1, *_1)$ and $(G_2, *_2)$ are two groups, then their product, denoted $(G_1 \times G_2, *)$ is simply $\{(g_1, g_2) : g_1 \in G_1, g_2 \in G_2\}$, where $(g_1, g_2) * (g_1', g_2')$ is $(g_1 *_1 g_1', g_2 *_2 g_2')$. The product of n groups, $G_1 \times G_2 \times \cdots \times G_n$ can be defined analogously; using this notation, the CRT can be stated in the language of group theory as follows.

Theorem A.13 (Chinese Remainder Version II). *If $m_0, m_1, \ldots, m_n$ are pairwise co-prime integers, then $\mathbb{Z}_{m_0 \cdot m_1 \cdots m_n} \cong \mathbb{Z}_{m_0} \times \mathbb{Z}_{m_1} \times \cdots \times \mathbb{Z}_{m_n}$.*

Problem A.14. *Prove theorem A.13*

A.1 Answers to selected problems

Problem A.6. $(a+1)^p \equiv_p \sum_{j=0}^{p} \binom{p}{j} a^{p-j} 1^j \equiv_p (a^p + 1) + \sum_{j=1}^{p-1} \binom{p}{j} a^{p-j}$. Note that $\binom{p}{j}$ is divisible by p for $1 \le j \le p - 1$, and so we have that $\sum_{j=1}^{p-1} \binom{p}{j} a^{p-j} \equiv_p 0$. Thus we can prove our claim by induction on a. The case $a = 1$ is trivial, and for the induction step we use the above observation to conclude that $(a+1)^p \equiv_p (a^p + 1)$ and we apply the induction hypothesis to get $a^p \equiv_p a$. Once we have prove $a^p \equiv_p a$ we are done since for a such that $\gcd(a, p) = 1$ we have an inverse a^{-1}, so we multiply both sides by it to obtain $a^{p-1} \equiv_p 1$.

Problem A.12. Construct the r in stages, so that at stage i it meets the first i congruences, that is, at stage i we have that $r \equiv r_j \pmod{m_j}$ for $j \in \{0, 1, \ldots, i\}$. Stage 1 is simple: just set $r \longleftarrow r_0$. Suppose the first i stages have been completed; let $r \longleftarrow r + (\Pi_{j=0}^{i} m_j)x$, where x satisfies $x \equiv (\Pi_{j=0}^{i} m_j)^{-1}(r_{i+1} - r) \pmod{m_{i+1}}$. We know that the inverse of $(\Pi_{j=0}^{i} m_j)$ exists (in $\mathbb{Z}_{m_{i+1}}$) since $\gcd(m_{i+1}, (\Pi_{j=0}^{i} m_i)) = 1$, and furthermore, this inverse can be obtained efficiently with the extended Euclid's algorithm.

A.2 Notes

For more algebraic background, see [Dummit and Foote (1991)] or [Alperin and Bell (1995)]. For number theory, especially related to cryptography, see [Hoffstein *et al.* (2008)]. A classical text in number theory is [Hardy and Wright (1980)].

Appendix B

Relations

In this section we present the basics of relations. Given two sets X, Y, $X \times Y$ denotes the set of (ordered) pairs $\{(x, y) | x \in X \wedge y \in Y\}$, and a *relation* R is just a subset of $X \times Y$, i.e., $R \subseteq X \times Y$. Thus, the elements of R are of the form (x, y) and we write $(x, y) \in R$ (we can also write xRy, Rxy or $R(x, y)$). In what follows we assume that we quantify over the set X and that $R \subseteq X \times X$; we say that

- R is *reflexive* if $\forall x, (x, x) \in R$,
- R is *symmetric* if $\forall x, y, (x, y) \in R$ if and only if $(y, x) \in R$,
- R is *antisymmetric* if $\forall x, y$, if $(x, y) \in R$ and $(y, x) \in R$ then $x = y$,
- R is *transitive* if $\forall x, y$, if $(x, y) \in R$ and $(y, z) \in R$ then $(x, z) \in R$.

Suppose that $R \subseteq X \times Y$ and $S \subseteq Y \times Z$. The *composition* of R and S is defined as follows:

$$R \circ S = \{(x, z) | \exists y, xRy \wedge ySz\}.$$

Let $R \subseteq X \times X$, then we define R^n recursively as follows:

$$R^0 = \mathrm{id}_{\mathrm{dom}(R)} := \{(x, x) \in R | x \in X\}, \qquad \text{(B.1)}$$

and $R^{i+1} = R^i \circ R$. Note that there are two different equalities in (B.1); "$=$" is the usual equality, and "$:=$" is a definition—that is, $\mathrm{id}_{\mathrm{dom}(R)}$ is defined to be $\{(x, x) \in R | x \in X\}$.

Theorem B.1. *The following three are equivalent:*

(1) R is transitive,
(2) $R^n \subseteq R$ for $n = 1, 2$,
(3) $\forall n \geq 1, R^n \subseteq R$.

There are two standard ways of representing *finite* relations, i.e., relations on $X \times Y$, where X and Y are finite sets. Let $X = \{a_1, \ldots, a_n\}$ and $Y = \{b_1, \ldots, b_m\}$, then we can represent a relation $R \subseteq X \times Y$:

(1) as matrices: then $M_R = (m_{ij})$ where:

$$m_{ij} = \begin{cases} 0 & (a_i, b_j) \in R \\ 1 & (a_i, b_j) \notin R \end{cases},$$

(2) and as directed graphs $G_R = (V_R, E_R)$: $V_R = X \cup Y$ and $a_i \bullet \longrightarrow \bullet b_j$ is an edge in E_R iff $(a_i, b_j) \in R$.

B.1 Closure

Let P be a property of relations, for example transitivity or symmetry. (We have already seen the concept of an abstract property in section 1.1.) Let $R \subseteq X \times X$ be a relation, with or without the property P. The relation S given by

$$S := \bigcap_{Q \text{ has P}, \ R \subseteq Q} Q, \qquad (B.2)$$

is called the *closure of R with respect to* P. Note that the closure may not exist, i.e., S could be $\emptyset$. Alternatively, S may be defined (equivalently) as the relation having the following three conditions:

(1) S has the property P,
(2) $R \subseteq S$, $\qquad$ (B.3)
(3) $\forall Q \subseteq X \times X$, "Q has P" and $R \subseteq Q$ implies that $S \subseteq Q$.

Problem B.2. *Prove that the two definitions are equivalent, that is, prove that S given by (B.2) and S given by (B.3) define the same relation.*

Theorem B.3. $\forall R,\ R \cup \text{id}_{\text{dom}(R)}$ *is the reflexive closure of R.*

Problem B.4. *Prove theorem B.3.*

Theorem B.5. $\forall R,\ R \cup R^{-1}$ *is the symmetric closure of R.*

Problem B.6. *Prove theorem B.5.*

Theorem B.7. $R^+ := \bigcup_{i=1}^{\infty} R^i$ *is the transitive closure of R.*

Proof. We check that R^+ has the three conditions given in (B.3). First that R^+ itself has the property (i.e., it is transitive):

$$xR^+y \wedge yR^+z \iff \exists m, n \geq 1, xR^my \wedge yR^mz$$
$$\implies \exists m, n \geq 1, x(R^m \circ R^n)z$$
$$\iff \exists m, n \geq 1, xR^{m+n}z$$
$$\iff xR^+z$$

so R^+ *is* transitive.

Second we check that $R \subseteq R^+$, but this follows by the definition of R^+.

We check now the last condition. Suppose S is transitive and $R \subseteq S$. Since S is transitive then $S^n \subseteq S$, for $n \geq 1$, i.e., $S^+ \subseteq S$, and since $R \subseteq S$, $R^+ \subseteq S^+$, so $R^+ \subseteq S$. $\square$

Theorem B.8. $R^* = \bigcup_{i=0}^{\infty} R^i$ *is the reflexive and transitive closure of* R.

Proof. $R^* = R^+ \cup \mathrm{id}_{\mathrm{dom}(R)}$. $\square$

B.2 Equivalence relation

Let X be a set, and let I be an index set. The family of sets $\{A_i | i \in I\}$ is called a *partition* of X iff

(1) $A_i \neq \emptyset$ for all i,
(2) $A_i \cap A_j = \emptyset$, for all $i \neq j$,
(3) $X = \bigcup_{i \in I} A_i$.

Note that $X = \bigcup_{x \in X} \{x\}$ is the *finest* partition possible, i.e., the set of all singletons. A relation $R \subseteq X \times X$ is called an *equivalence relation* iff

(1) R is reflexive,
(2) R is symmetric,
(3) R is transitive.

For example, if x, y are strings over $\{0, 1\}^*$, then the relation given by $R = \{(x, y) | \mathrm{length}(x) = \mathrm{length}(y)\}$ is an equivalence relation. Another example is $xRy \iff x = y$, i.e., the equality relation is the equivalence relation *par excellence*. Yet another example: $R = \{(a, b) | a \equiv b \pmod{m}\}$ is an equivalence relation (where "$\equiv$" is the congruence relation defined on page 114).

Theorem B.9. *Let* $F : X \longrightarrow X$ *be any* total *function (i.e., a function defined on all its inputs). Then the relation R on X defined as: $xRy \iff F(x) = F(y)$, is an equivalence relation.*

Problem B.10. *Prove theorem B.9.*

Let R be an equivalence relation on X. For every $x \in X$, the set $[x]_R = \{y | xRy\}$ is the *equivalence class of x with respect to R.*

Theorem B.11. *Let $R \subseteq X \times X$ be an equivalence relation. The following are equivalent:*

(1) aRb
(2) $[a] = [b]$
(3) $[a] \cap [b] \neq \emptyset$

Proof. $(1) \Rightarrow (2)$ Let aRb. Suppose $c \in [a]$. Then aRc, so cRa (by symmetry). Since $cRa \wedge aRb$, cRb (transitivity), so bRc (symmetry), so $c \in [b]$. Hence $[a] \subseteq [b]$, and similarly $[b] \subseteq [c]$.
    $(2) \Rightarrow (3)$ Obvious, since $a \in [a]$.
    $(3) \Rightarrow (1)$ Let $c \in [a] \cap [b]$, so aRc and bRc, so by symmetry $aRc \wedge cRb$, so by transitivity aRb.   □

Corollary B.12. *If R is an equivalence, then $(a,b) \notin R$ iff $[a] \cap [b] = \emptyset$.*

For every equivalence relation $R \subseteq X \times X$, let X/R denote the set of all equivalence classes of R.

Theorem B.13. *X/R is a partition of X.*

Proof.   The only thing that we need to prove is that $X = \bigcup_{A \in X/R} A$, so take $a \in X$, clearly $a \in [a] \in X/R$, so we are done.   □

Let R_1, R_2 be equivalence relations. R_1 is called a *refinement* of R_2 if $R_1 \subseteq R_2$.

Lemma B.14. *If R_1 is a refinement of R_2 then for all $a \in X$,*

(1) $[a]_{R_1} \subseteq [a]_{R_2}$
(2) $[a]_{R_1} \cap [a]_{R_2} \neq \emptyset \Rightarrow [a]_{R_1} \subseteq [a]_{R_2}$.

If X/R is finite then $\mathrm{index}(R) := |X/R|$, i.e., the *index* of R (in X) is the size of X/R.

Theorem B.15. *$R_1 \subseteq R_2 \Rightarrow \mathrm{index}(R_1) \geq \mathrm{index}(R_2)$.*

Problem B.16. *Prove theorem B.15.*

B.3   Partial orders

A relation $R \subseteq X \times X$ is called a *partial order* (a *poset* for short) iff it is

(1) reflexive
(2) antisymmetric
(3) transitive

We frequently use notation like $(X, \preceq), (X, \sqsubseteq), (X, \leq)$, instead of (X, R). An alternative definition is as follows: a relation $\prec \subseteq X \times X$ is a *sharp partial order* iff

(1) $x \prec y \Rightarrow \neg(y \prec x)$
(2) $x \prec y \wedge y \prec z \Rightarrow x \prec z$

Theorem B.17. *A relation $\preceq$ defined as $x \preceq y \iff x \prec y \vee x = y$ is a partial order.*

Let $(X, \preceq)$ be a poset. We say that x, y are *comparable* if $x \preceq y$ or $y \preceq x$. Otherwise, they are *incomparable*. Let $x \sim y$ be short for x, y are incomparable, i.e., $x \sim y \iff \neg(x \preceq y) \wedge \neg(y \preceq x)$. In general, for every pair x, y exactly one of the following is true

$$x \prec y, \quad y \prec x, \quad x = y, \quad x \sim y$$

A poset $(X, \preceq)$ is *total* or *linear* iff all x, y are comparable, i.e., $\sim = \emptyset$.

Some examples of posets: if X is a set, then $(\mathcal{P}(X), \subseteq)$ is a poset. Let $\mathbb{Z}^+$ be the set of positive integers, and let $a|b$ be the "a divides b" relation (that we define on page 113). Then, $(\mathbb{Z}^+, |)$ is a poset.

If $(X_1, \preceq_1), (X_2, \preceq_2)$ are two posets, then the *component-wise order* is $(X_1 \times X_2, \preceq_C)$ defined as follows: $(x_1, x_2) \preceq (y_1, y_2) \iff x_1 \preceq_1 y_1 \wedge x_2 \preceq_2 y_2$, and it is also a poset. The *lexicographic order* $(X_1 \times X_2, \preceq_L)$ is defined as follows: $(x_1, x_2) \preceq_L (y_1, y_2) \iff (x_1 \prec_1 y_1) \vee (x_1 = y_1 \wedge x_2 \preceq_2 y_2)$. Finally, $(X, \preceq)$ is a *stratified order* iff $(X, \preceq)$ is a poset, and furthermore $(x \sim y \wedge y \sim z) \Rightarrow (x \sim z \vee x = z)$. Define $a \approx b \iff a \sim b \vee a = b$.

Theorem B.18. $(X, \preceq)$ *is a stratified order iff* $\approx = \sim \cup \mathrm{id}_X$ *is an equivalence relation.*

In mathematics nomenclature can be the readers greatest enemy. The string of symbols "$\approx = \sim \cup \mathrm{id}_X$" is a great example of obfuscation; how to make sense of it? Yes, it is very succinct, but it takes practice to be able to read it. What we are saying here is that the order we called $\approx$ is actually equal to the order that we obtain by taking the union of the order $\sim$ and id_X.

Theorem B.19. $(X, \preceq)$ *is a stratified order iff there exists a total order* $(T, \preceq_T)$ *and a function* $f : X \longrightarrow T$ *such that* f *is onto and* $a \preceq b \iff f(a) \preceq_T f(b)$.

The element $a \in A$ is a *least element* of A if for all $x \in A$, $a \preceq x$. $(X, \preceq)$ is a *well-ordered* set if it is a total order and for every $A \subseteq X$, such that $A \neq \emptyset$, A has a least element.

Let $(X, \preceq)$ be a poset, and let $A \subseteq X$ be a subset; some more definitions:

- $a \in X$ is *minimal* in X iff $\forall x \in X, \neg(x \prec a)$.
- $a \in X$ is *maximal* in X iff $\forall x \in X, \neg(a \prec x)$.
- $a \in X$ is the *greatest element* in X iff $\forall x \in X, x \preceq a$.
- $a \in X$ is an *upper bound* of A iff $\forall x \in A, x \preceq a$.
- $a \in X$ is a *lower bound* if A iff $\forall x \in A, a \preceq x$.
- $a \in X$ is the *least upper bound (supremum)* of A, denoted $\sup(A)$ iff
  - $\forall x \in A, x \preceq a$
  - $\forall b \in X, (\forall x \in A, x \preceq b) \Rightarrow a \preceq b$
- $a \in X$ is the *greatest lower bound (infimum)* of A, denoted $\inf(A)$ iff
  - $\forall x \in A, a \preceq x$
  - $\forall b \in X, (\forall x \in A, b \preceq x) \Rightarrow b \preceq a$

It is important to note that $\sup(A), \inf(A)$ may or may not exist.

A poset is *dense* if $\forall x, y$ if $x < y$, then $\exists z, x < z < y$. For example, $(\mathbb{R}, \leq)$ is a total dense order.

B.4 Lattices

A poset $(X, \preceq)$ is a *lattice* iff $\forall a, b \in X$, $\inf(\{a, b\})$ and $\sup(\{a, b\})$ exists (in X). For example, $(\mathcal{P}(X), \subseteq)$ is a lattice for every X. Also, every total order is a lattice. We use the following notation: $a \sqcup b := \sup(\{a, b\})$ and $a \sqcap b := \inf(\{a, b\})$.

Theorem B.20. *Let* $(X, \preceq)$ *be a lattice. Then,* $\forall a, b \in X$,

$$a \preceq b \iff a \sqcap b = a \iff a \sqcup b = b.$$

Theorem B.21. *Let* $(X, \preceq)$ *be a lattice. Then, the following hold for all* $a, b, c \in X$:

(1) $a \sqcup b = b \sqcup a$ and $a \sqcap b = b \sqcap a$ (commutativity)
(2) $a \sqcup (b \sqcup c) = (a \sqcup b) \sqcup c$ and $a \sqcap (b \sqcap c) = (a \sqcap b) \sqcap c$ (associativity)
(3) $a \sqcup a = a$ and $a \sqcap a = a$ (idempotence)
(4) $a = a \sqcup (a \sqcap b)$ and $a = a \sqcap (a \sqcup b)$ (absorption)

Theorem B.22. *For the lattice $(\mathcal{P}(X), \subseteq)$ we have: $A \sqcup B = A \cup B$ and $A \sqcap B = A \cap B$.*

A lattice $(X, \preceq)$ is *complete* iff $\forall A \subseteq X$, $\sup(A), \inf(A)$ both exist. We denote $\bot = \inf(X)$ and $\top = \sup(X)$.

Theorem B.23. *$(\mathcal{P}(X), \subseteq)$ is a complete lattice, and the following hold $\forall \mathcal{A} \subseteq \mathcal{P}(X)$, $\sup(\mathcal{A}) = \bigcup_{A \in \mathcal{A}} A$ and $\inf(\mathcal{A}) = \bigcap_{A \in \mathcal{A}} A$, and $\bot = \emptyset$ and $\top = X$.*

Problem B.24. *Prove theorem B.23.*

Theorem B.25. *Every finite lattice is complete.*

Proof. Let $A = \{a_1, \ldots, a_n\}$. Define $b = a_1 \sqcap \ldots \sqcap a_n$ (with parenthesis associated to the right). Then $b = \inf(A)$. Same idea for the supremum. $\square$

B.5 Fixed point theory

Consider the equation $\vec{x} = F(\vec{x})$. A solution $\vec{a}$ of this equation is a *fixed point of F*.

Let $(X, \preceq)$ and $(Y, \sqsubseteq)$ be two posets. A function $f : X \longrightarrow Y$ is *monotone* iff $\forall x, y \in X$, $x \preceq y \Rightarrow f(x) \sqsubseteq f(y)$. For example, $f_B : \mathcal{P}(X) \longrightarrow \mathcal{P}(X)$, where $B \subseteq X$, defined as $\forall x \subseteq X$, $f_B(x) = B - x$ is *not* monotone. On the other hand, $g_B(x) = B \cup x$ and $h_B(x) = B \cap x$ are both monotone.

Let $(X, \preceq)$ be a poset, and let $f : X \longrightarrow X$. A value $x_0 \in X$ such that $x_0 = f(x_0)$ is called a *fixed point* of f. A fixed point may not exist (see example f_B in the above paragraph), and there may be many fixed points; for example, $f(x) = x$ has $|X|$ many fixed points.

Theorem B.26 (Knaster-Tarski (1)). *Let $(X, \preceq)$ be a complete lattice, and let $f : X \longrightarrow X$ be a monotone function. Then the least fixed point of the equation $x = f(x)$ exists and it is equal to $\inf(\{x | f(x) \preceq x\})$.*

Proof. Let $x_0 = \inf(\{x | f(x) \preceq x\})$. First we show that $x_0 = f(x_0)$. Let $B = \{x | f(x) \preceq x\}$, and note that $B \neq \emptyset$ because $\top = \sup(X) \in B$. Let $x \in B$, so we have $x_0 \preceq x$, hence since f is monotone, $f(x_0) \preceq f(x)$, i.e.,

$$f(x_0) \preceq f(x) \preceq x.$$

This is true for each x in B, so $f(x_0)$ is a lower bound for B, and since x_0 is the greatest lower bound of B, it follows that $f(x_0) \preceq x_0$.

Since f is monotone it follows that $f(f(x_0)) \preceq f(x_0)$, which means that $f(x_0)$ is in B. But then $x_0 \preceq f(x_0)$, which means that $x_0 = f(x_0)$.

It remains to show that x_0 is the least fixed point. Let $x' = f(x')$. This means that $f(x') \preceq x'$, i.e., $x' \in B$. But then $x_0 \preceq x'$. $\square$

Theorem B.27 (Knaster-Tarski (2)). *Let $(X, \preceq)$ be a complete lattice, and let $f : X \longrightarrow X$ be a monotone function. Then the greatest fixed point of the equation $x = f(x)$ exists and it is equal to $\sup(\{x | f(x) \preceq x\})$.*

These theorems are not constructive, but the situation is good for finite sets.

Theorem B.28 (Knaster-Tarski for finite sets). *Let $(X, \preceq)$ be a lattice, $|X| = m$, $f : X \longrightarrow X$ a monotone function. Then $f^m(\bot)$ is the least fixed point, and $f^m(\top)$ is the greatest fixed point.*

Proof. Since $|X| = m$, $(X, \preceq)$ is a complete lattice, $\bot = \inf(X)$ and $\top = \sup(X)$ both exist. Since f is monotone, and $\bot \preceq f(\bot)$, we have $f(\bot) \preceq f(f(\bot))$, i.e., $f(\bot) \preceq f^2(\bot)$. Continuing to apply monotonicity we obtain:

$$f^0(\bot) = \bot \preceq f(\bot) \preceq f^2(\bot) \preceq f^3(\bot) \preceq \cdots \preceq f^i(\bot) \preceq f^{i+1}(\bot) \preceq \cdots.$$

Consider the above sequence up to $f^m(\bot)$. It has length $(m + 1)$, but X has only m elements, so there are $i < j$, such that $f^i(\bot) = f^j(\bot)$. Since $\preceq$ is an order, it follows that

$$f^i(\bot) = f^{i+1}(\bot) = \cdots = f^j(\bot),$$

so $x_0 = f^i(\bot)$ is a fixed point as

$$f(x_0) = f(f^i(\bot)) = f^{i+1}(\bot) = f^i(\bot) = x_0.$$

Clearly $f^{j+1}(\bot) = f(f^j(\bot)) = f(x_0) = x_0$, so in fact $\forall k \geq i$, $x_0 = f^k(\bot)$, and so $f^m(\bot) = x_0$, so $f^m(\bot)$ is a fixed point.

We now suppose that x is another fixed point of f, i.e., $x = f(x)$. Since $\bot \preceq x$, and f is monotone, we conclude $f(\bot) \preceq f(x) = x$, i.e., $f(\bot) \preceq x$.

Again, since f is monotone, $f(f(\perp)) \preceq f(x) = x$, so $f^2(\perp) \preceq x$. Hence, repeating this procedure sufficiently many times, we obtain $f^i(\perp) \preceq x$ for each i, so we get $x_0 = f^m(\perp) \preceq x$.

We do a similar argument for "greatest." $\qquad\square$

The situation is even better for the standard lattice $(\mathcal{P}(X), \subseteq)$, if X is finite.

Theorem B.29. *Let X be a finite set, $|X| = n$, $f : \mathcal{P}(X) \longrightarrow \mathcal{P}(X)$ is monotone. Then $f^{n+1}(\emptyset)$ is the least fixed point, and $f^{n+1}(X)$ is the greatest fixed point.*

Proof. Note that the previous theorem says that $f^{2^n}(\emptyset)$ is the least fixed point, and $f^{2^n}(X)$ is the greatest fixed point, since $|\mathcal{P}(X)| = 2^n$, $\perp = \emptyset$ and $\top = X$, for the lattice $(\mathcal{P}(X), \subseteq)$. But this theorem claims $(n+1)$ instead of 2^n. The reason is that $\emptyset \subseteq f(\emptyset) \subseteq f^2(\emptyset) \subseteq \cdots \subseteq f^{n+1}(\emptyset)$ *must* have two repeating sets (because $|X| = n$). $\qquad\square$

Let $(X, \preceq)$ be a complete lattice. A function $f : X \longrightarrow X$ is called

- *upward continuous* iff $\forall A \subseteq X$, $f(\sup(A)) = \sup(f(A))$,
- *downward continuous* iff $\forall A \subseteq X$, $f(\inf(A)) = \inf(f(A))$,
- *continuous* if it is both upward and downward continuous.

Lemma B.30. *If $f : X \longrightarrow X$ is upward (downward) continuous, then it is monotone.*

Proof. Let f be upward continuous and $x \preceq y$, so $x = \inf(\{x, y\})$ and $y = \sup(\{x, y\})$, and

$$f(x) \preceq \sup(\{f(x), f(y)\}) = \sup(f(\{x, y\})) = f(\sup(\{x, y\})) = f(y).$$

A similar argument for downward continuous. $\qquad\square$

Theorem B.31 (Kleene). *If $(X, \preceq)$ is a complete lattice, $f : X \longrightarrow X$ is an upward continuous function, then $x_0 = \sup(\{f^n(\perp)|n = 1, 2, \ldots\})$ is the least fixed point of f.*

Proof. Note that $\perp \preceq f(\perp)$, so by monotonicity of f, we have that

$$\perp \preceq f(\perp) \preceq f^2(\perp) \preceq f^3(\perp) \preceq \cdots \qquad (\text{B.4})$$

and,

$$f(x_0) = f(\sup(\{f^n(\perp)|n = 1, 2, \ldots\}))$$

and since f is upward continuous

$$= \sup(f(\{f^n(\bot)|n = 1, 2, \ldots\}))$$
$$= \sup(\{f^{n+1}(\bot)|n = 1, 2, \ldots\})$$

and by (B.4),

$$= \sup(\{f^n(\bot)|n = 1, 2, \ldots\}) = x_0$$

so $f(x_0) = x_0$, i.e., x_0 is a fixed point.

Let $x = f(x)$. We have $\bot \preceq x$ and f is monotone, so $f(\bot) \preceq f(x) = x$, i.e., $f(\bot) \preceq x$, $f^2(\bot) \preceq f(x) = x$, etc., i.e., $f^n(\bot) \preceq x$, for all n, so by the definition of sup,

$$x_0 = \sup(\{f^n(\bot)|n = 1, 2, \ldots\}) \preceq x,$$

so x_0 is the least fixed point.   □

B.6   Answers to selected problems

Problem B.10. R is reflexive since $F(x) = F(x)$; R is symmetric since $F(x) = F(y)$ implies $F(y) = F(x)$ (equality is a symmetric relation); R is transitive because $F(x) = F(y)$ and $F(y) = F(z)$ implies $F(x) = F(z)$ (again by transitivity of equality).

B.7   Notes

This chapter is based on hand-written lecture slides of Ryszard Janicki.

Bibliography

Agrawal, M., Kayal, N. and Saxena, N. (2004). Primes is in P, *Annals of Mathematics* **160**, 2, pp. 781–793.

Alford, W. R., Granville, A. and Pomerance, C. (1994). There are infinitely many Carmichael numbers, *Annals of Mathematics* **139**, 3, pp. 703–722.

Alperin, J. L. and Bell, R. B. (1995). *Groups and Representations* (Springer).

Berkowitz, S. J. (1984). On computing the determinant in small parallel time using a small number of processors, *Information Processing Letters* **18**, 3, pp. 147–150.

Borodin, A. and El-Yaniv, R. (1998). *Online Computation and Competitive Analysis* (Cambridge University Press).

Cenzer, D. and Remmel, J. B. (2001). Proof-theoretic strength of the stable marriage theorem and other problems, *Reverse Mathematics* , pp. 67–103.

Cormen, T. H., Leiserson, C. E., Rivest, R. L. and Stein, C. (2001). *Introduction to Algorithms* (McGraw-Hill Book Company), second Edition.

Delfs, H. and Knebl, H. (2007). *Introduction to Cryptography* (Springer).

Dijkstra, E. W. (1989). On the cruelty of really teaching computing science, *Communications of the ACM* **32**, 12.

Dummit, D. S. and Foote, R. M. (1991). *Abstract Algebra* (Prentice Hall).

Engel, A. (1998). *Problem-Solving Strategies* (Springer).

Halmos, P. R. (1995). *Linear algebra problem book* (The mathematical association of America).

Hardy, G. H. and Wright, E. M. (1980). *An Introduction to the Theory of Numbers*, 5th edn. (Oxford University Press).

Harel, D. (1987). *Algorithmics: The Spirit of Computing* (The Addison-Wesley Publishing Company), ISBN 0-201-19240-3.

Hoffman, P. (1998). *The Man Who Loved Only Numbers: The Story of Paul Erdős and the Search for Mathematical Truth* (Hyperion).

Hoffstein, J., Pipher, J. and Silverman, J. H. (2008). *An Introduction to Mathematical Cryptography* (Springer).

Karp, R. M. and Rabin, M. O. (1987). Efficient randomized pattern-matching algorithms. *IBM Journal of Research and Development* **31**, 2, pp. 249–260.

Kleinberg, J. and Tardos, É. (2006). *Algorithm Design* (Pearson Education).

Knuth, D. E. (1997). *The Art of Computer Programming*, Vol. 1, Fundamental Algorithms, 3rd edn. (Addison Wesley).

Kozen, D. (2006). *Theory of Computation* (Springer).

Manna, Z. (1974). *Mathematical Theory of Computation* (McGraw-Hill Computer Science Series).

Papadimitriou, C. H. (1994). *Computational Complexity* (Addison-Wesley).

Press, W. H., Vetterling, W. T., Teukolsky, S. A. and Flannery, B. P. (2007). *Numerical Recipes: The Art of Scientifc Computing*, 3rd edn. (Cambridge University Press).

Rosen, K. H. (2007). *Discrete Mathematics and Its Applications*, 6th edn. (McGraw Hill).

Shustek, L. (2009). Interview, *Communications of the ACM* **52**, 3, pp. 38–41.

Singh, S. (1999). *The Code Bok: The evolution of secrecy, from Mary, Queen of Scots, to Quantum Cryptography* (Doubleday).

Sipser, M. (2006). *Introduction to the Theory of Computation* (Thompson), second Edition.

Solovay, R. and Strassen, V. (1977). A fast monte-carlo test for primality, *SIAM Journal of Computing* **6**, pp. 84–86.

Soltys, M. (2002). Berkowitz's algorithm and clow sequences, *Electronic Journal of Linear Algebra* **9**, pp. 42–54.

Soltys, M. (2009). *An Introduction to Computational Complexity* (Jagiellonian University Press), to appear.

van Vliet, H. (2000). *Software Engineering: Principles and Practice*, 2nd edn. (Wiley).

Velleman, D. J. (2006). *How To Prove It*, 2nd edn. (Cambridge University Press).

von zur Gathen, J. and Gerhard, J. (1999). *Modern computer algebra* (Cambridge University Press).

Zingaro, D. (2008). *Invariants: A generative approach to programming* (College Publications).

Index

activity selection, *59–61*
algorithm
  $F(x, y)$, 12
  c-competitive, 78
  $f_1(x, y)$, 13
  $f_2(x, y)$, 13
  $f_3(x, y)$, 13
  Activity Selection, 70
  approximation, 57
  Berkowitz's, 97, 111
  CYK, 66
  Dispersed Knapsack
    Problem 4.12, 58
    Solution to problem 4.12, 69
  Division, 6
  Euclid, 8
  extended Euclid's, **8**, 107, 108, 114,
    118
  Floyd, 54
  Gale-Shapley, 16
  Gram-Schmidt, 10
  Job Scheduling, 34
  Kruskal, 31
  Las Vegas, 101
  Longest Monotone Subsequence
    (LMS), 52
  Make Change, 37
  marking, *see* marking algorithms
  Merge Two Lists, 44
  Mergesort, 44
  Merging Components, 31
  Monte Carlo, 97

offline
  Longest Forward Distance
    (LFD), 79, 88
online
  First-In/First-Out (FIFO), 82
  Flush When Full (FWF), 87
  Least Recently Used (LRU),
    82
  Move To Front (MTF), 75
Palindromes, 9
pattern matching, 101
Perfect Matching, 97
Powers of 2, 23
Problem 4.21, 64
Problem 1.21, 10
Problem 1.22, 10
Problem 1.25, 26
Rabin-Miller, 103
recursive, 12
Recursive binary multiplication, 47
Savitch, 49
Shank's Babystep-Giantstep, 107
Simple Knapsack (SKS), 56
Solution to problem 2.21, 41
Ulam's, 11
all pairs shortest path, *52–54*
alphabet, 64
amortized cost, 76
amplification, 99
average-case, 78

Belady's anomaly, 82

big-Oh notation, ix
brute force, 4

Chinese Remainder Theorem, *see*
  theorem
CIP, *see* Complete Induction
  Principle
Collatz conjecture, *see* Ulam,
  Stanisław—$3x + 1$ algorithm
competitive analysis, 78
competitive ratio, 78
Complete Induction Principle, 2
congruence, 114
consecutive subsequence sum, *63–64*
context free grammar, **64**, *64–66*
  derivation, 65
  language of, 65
  terminal, 64
Cook, Stephen, ix
correctness of algorithms, *5–14*
  correctness, 5
  partial, 5
  post-condition, 5
  pre-condition, 5
CRT, *see* theorem

demand paging, *79–82*
DHP, *see* Diffie-Hellman Problem
  (DHP)
Diffie-Hellman key exchange, *see*
  public key cryptography
Diffie-Hellman Problem (DHP), 106
discrete log problem (DLP), 106
distributional, 78
divides, 113
divisor, 113
DLP, *see* discrete log problem (DLP)
dynamic list accessing model, 78

ElGamal, *see* public key
  cryptography

factor, 113
Fibonacci sequence, **2**, 18, 100
FIFO, *see* fist in first out (FIFO)
fingerprinting, 101

first in first out (FIFO), *82*
flush when full (FWF), *87–88*
free transposition, 74
function
  ceil, ix
  domain of definition, 12
  Euler totient, 114
  fixed point, 13, 125
  floor, ix
  total, 12, **122**
FWF, *see* flush when full (FWF)

general knapsack problem, *58–59*
graph
  acyclic, 29
  connected, 29
  connected component, 30
  cycle, 29
  digraph, 53
  directed, 53
  path, 29
  spanning forest, 32
  spanning tree, 29
  tree, 29
    infix, 3
    leaf, 30
    minimum cost spanning
      (MCST), 30
    postfix, 3
    prefix, 3
  undirected, 29
  vertex
    degree, 29
greatest common divisor, 7
group, **115**, *113–118*
  $(\mathbb{Q}, +)$, 115
  $(\mathbb{Z}_p, +)$, 115
  $(\mathbb{Z}_p^*, \cdot)$, 115
  $GL(n, \mathbb{F})$, 115
  Abelian, 115
  associative law, 115
  closure, 115
  commutative law, 115
  homomorphism, 117
  identity law, 115
  inverse law, 115

isomorphism, 117
  left coset, 116
  order, 116
  subgroup, 116
  symmetric, 115
group theory, *see* group

Hasse's Algorithm, *see* Ulam,
  Stanisław—$3x + 1$ algorithm
hit, 78
Huffman codes, *42*

Induction, *1–3*
  Principle (IP), 1
intractable, 54
Invariance, *4–5*
  Technique (IT), 4
IP, *see* Induction Principle
IT, *see* Invariance Technique

Janicki, Ryszard, 128
jobs with deadlines and profits, *34–36*
jobs with deadlines, durations and
  profits, *62–63*
Joyce, James, 28

Kukatani's Problem, *see* Ulam,
  Stanisław—$3x + 1$ algorithm

Lagrange formula, 96
lattice, 124
  absorption, 125
  associativity, 125
  commutativity, 125
  complete, 125
  idempotence, 125
Least Number Principle (LNP), 3
least recently used (LRU), *82–85*
lemma
  Exchange, 33
  Schwarz-Zippel, 97
LFD, *see* longest forward distance
  (LFD)
list accessing, *74–78*
LMS, *see* Longest Monotone
  Subsequence

LNP, *see* Least Number Principle
longest forward distance (LFD),
  88–91
Longest Monotone Subsequence,
  51–52
loop invariant, 6
LRU, *see* least recently used (LRU)

Manhattan Project, 111
marking algorithms, *86–87*
marriage problem, 14
  blocking pair, 15
  boy-optimal, 17
  boy-pessimal, 17
  feasible, 17
  unstable, 15
MCST, *see* graph, minimum cost
  spanning tree
mergesort, *44–45*
minimum cost spanning tree, *29–34*
miss, 79
modus ponens, 1
Monte Carlo, *see* algorithm
multiplying numbers in binary, *45–48*

NP-hard, 54
number theory, *113–118*

optimal offline algorithm, 78

page fault, 79
page fault model, 79
pages, 78
paging, *78–91*
paid transposition, 74
palindrome, 8
partial correctness, *see* correctness of
  algorithms
partial order, 123
  comparable, 123
  component-wise, 123
  dense, 124
  function
    continuous, 127
    downward continuous, 127
    fixed point, 125

monotone, 125
    upward continuous, 127
greatest element, 124
greatest lower bound, 124
incomparable, 123
infimum, 124
least element, 124
least upper bound, 124
lexicographic, 123
linear, 123
lower bound, 124
maximal, 124
minimal, 124
sharp, 123
stratified, 123
supremum, 124
total, 123
upper bound, 124
well-ordered, 124
pattern matching, *100–101*
perfect matching, *96–100*
phases, 86
    k-phase partition, 86
PKC, *see* public key cryptosystem
    (PKC)
poset, *see* partial order
post-condition, *see* correctness of
    algorithms
potential function, 75
pre-condition, *see* correctness of
    algorithms
primality testing, *101–105*
prime number, 113
program refinement, 56
promise problem, 58
public key cryptography, *105–109*
    Diffie-Hellman key exchange, *106*
    ElGamal, *107*
    RSA, *107–110*
public key cryptosystem (PKC), 105
    ciphertext, 105
    decryption, 105
    encryption, 105
    keys, 105
    private key, 105
    secret key, 105

trapdoor, 105

quotient, 6

recurrence
    Master method, 50
reduction, **62**, 99
regularity condition, 50
relation, **119**, *119–128*
    antisymmetric, 119
    closure, 120
    composition, 119
    directed graph representation, 120
    equivalence, 121
        class, 122
    matrix representation, 120
    partition, 121
        finest, 121
    refinement, 122
    reflexive, 119
    symmetric, 119
    transitive, 119
reminder, 6
RSA, *see* public key cryptography

Savitch's algorithm, *48–49*
Savitch, Walter, 48
    algorithm, *see* algorithm
simple knapsack problem, *54–58*
    greedy attempt, 57, 68
SKS, *see* simple knapsack problem,
    algorithm
stable marriage, *14–17*
static list accessing model, 75
Syracuse Problem, *see* Ulam,
    Stanisław—3*x* + 1 algorithm

Teller, Edward, 111
termination, 5
theorem
    Chinese Remainder (CRT), 117
    Chinese Remainder Version II, 118
    Euler, 111, **117**
    Euler's, 117
    Fermat's Little, 102, 103, 111, **114**,
        115, 117

Fundamental of Algebra, 98
Fundamental of Arithmetic, 18,
   113
Kleene, 127
Knaster-Tarski (1), 125
Knaster-Tarski (2), 126
Knaster-Tarski for finite sets, 126
Lagrange, 116
Prime Number, 101, 105
Turing machine, 54
   non-deterministic, 54

Ulam, Stanisław
   $3x + 1$ algorithm, **11**
   Monte Carlo method, 111

virtual memory system, 78
von Neumann, John, 111

worst case, 78